"Engaging. . . . Friedman's usual off-kilter charm prevails throughout."   —*Publishers Weekly*

"Great stuff. . . . Flashes of brilliance and laugh-out-loud observations."   —*Denver Rocky Mountain News*

PRAISE FOR KINKY FRIEDMAN

"Kinky, Mozart, Shakespeare—with what could I equal them?"
—Joseph Heller

"Kinky is a hip hybrid of Groucho Marx and Sam Spade."
—*Chicago Tribune*

"A Texas legend."   —George W. Bush

"We read [Kinky] to find out how far he will go. . . . Nothing is sacred in a Kinky Friedman book—therein lies his charm."
—*Washington Post Book World*

"Kinky Friedman is a true American original, something it's not easy to be in today's carbon-copy culture."   —Steve Allen

"The new Mark Twain."   —*Southern Living*

"The world's funniest, bawdiest, and most politically incorrect country music singer-turned-writer."   —*New York Times*

"Sometimes he's just outrageous. Most of the time, he's outrageously funny."   —*People*

Briana Voelkel

About the Author

KINKY FRIEDMAN is the author of *Kinky Friedman's Guide to Texas Etiquette*, the novel *Kill Two Birds & Get Stoned*, as well as eighteen mysteries whose main character is Kinky Friedman. He lives with five dogs, three donkeys, one armadillo, a million hummingbirds, and a much-used Smith Corona typewriter on a ranch in the Texas Hill Country. With any luck, he'll be the next governor of the state of Texas.

ALSO BY KINKY FRIEDMAN

Greenwich Killing Time

A Case of Lone Star

When the Cat's Away

Frequent Flyer

Musical Chairs

Elvis, Jesus & Coca-Cola

Armadillos & Old Lace

God Bless John Wayne

The Love Song of J. Edgar Hoover

Roadkill

Blast from the Past

Spanking Watson

The Mile High Club

Kinky Friedman's Guide to Texas Etiquette

Kill Two Birds & Get Stoned

Meanwhile Back at the Ranch

Curse of the Missing Puppethead

The Prisoner of Vandam Street

KINKY FRIEDMAN

'SCUSE Me
WHILE I WHIP
This OUT

Reflections on
Country Singers, Presidents,
and Other Troublemakers

Harper

An Imprint of HarperCollins*Publishers*

A hardcover edition of this book was published in 2004 by William Morrow, an imprint of HarperCollins Publishers.

FIRST HARPER PAPERBACK PUBLISHED 2005.

Designed by Renato Stanisic

Library of Congress Cataloging-in-Publication Data has been applied for.

ISBN 0-06-053975-5
ISBN-10: 0-06-053976-3 (pbk.)
ISBN-13: 978-0-06-053976-4 (pbk.)

05 06 07 08 09 ❖/RRD 10 9 8 7 6 5 4 3 2 1

This one's for The Navigator

CONTENTS

PART 3

OTHER TROUBLEMAKERS

PART 4

ALL OF THE ABOVE

Introduction

Both *Willie Nelson and George W. Bush are* friends of mine. In fact, over the years I've known them, to paraphrase my father, they've moved up from friends to contacts. Willie, depending on which polls you read, is quite often more popular than George, but then Willie never gets to ride on Air Force One.

In *'Scuse Me While I Whip This Out,* you may see another, extremely humorous side of these two great Americans, a side that most people never get to observe. After all, I'm a writer of fiction who tells the truth. That's one reason I'm contemplating running for (or, at my age, taking a leisurely stroll toward), governor of Texas in 2006. I'm fifty-nine but I read at the sixty-one-year-old level.

Willie and George, of course, have both offered to help with my campaign. George told me he would answer any questions that I had. "I'll be your one-man focus group," he told me recently. Willie says, "In a Friedman administration, I'd like to be head of the Texas Rangers. If not that, then head of the DEA." George, as you will soon discover, even went so far as to offer me a job. Sort of. Willie, not to be outdone, has offered me many things. One of them was a joint the size of a large kosher salami.

In the course of this book, as well as George and Willie, you'll also meet other endearing characters I have known, such as Bill Clinton, Bob Dylan, Joe Heller, Don Imus, Irv Rubin, and Billy Joe Shaver, as well as those I have known only in spirit, Moses, Jesus, Jack Ruby, and Hank Williams to name a few. As the Beatles sang, "Some are dead and some are living / In my life I've loved them all."

The people in this book are troublemakers because they have caused us to think, and quite often they have led us in new and different directions. They have dared to stir the putrid pot of humanity every now and then, and each one of them has paid dearly for it. I feel proud and privileged to live in a country where I can so freely follow in their footsteps.

PART 1

Country Singers

Outlaws

The life of a country singer can at times be very tedious. You have to pretend that your life is a financial pleasure even when your autographs are bouncing. You often fall prey to the serious songwriters' self-pity syndrome. You begin to believe that all dentists and married couples are happier than you are. Many's the night you feel

> **SUDDENLY YOU FIND YOU'RE A JET-SET GYPSY CRYIN' ON THE SHOULDER OF THE HIGHWAY.**

lonely, empty, homesick for heaven. Everybody you know thinks you've got it made and suddenly you find you're a jet-set gypsy cryin' on the shoulder of the highway. Believe me when I tell you, it's lonely in the middle.

But long before the Outlaw Movement, as we now call it, came along in the 1970s, there were great voices in country music who never fit in wherever they were. Their spirits and songs somehow survived that all-pervasive white noise called the Nashville Sound even before it had a name. I

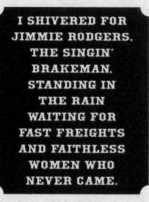

I SHIVERED FOR JIMMIE RODGERS, THE SINGIN' BRAKEMAN, STANDING IN THE RAIN WAITING FOR FAST FREIGHTS AND FAITHLESS WOMEN WHO NEVER CAME.

shivered for Jimmie Rodgers, the Singin' Brakeman, standing in the rain waiting for fast freights and faithless women who never came, who finally sang the TB blues, dying out like a train whistle in the night, the lantern still swinging in his hand. And Hank Williams, skinny, hungry, spiritually horny, for whom all the world was a stage. Shakespeare of the sequined summer stock. Hank died when he was twenty-nine years old—perfect timing for a country music legend, dreaming his last backseat dreams in the backseat of that shimmering, earthbound Cadillac, on his way to a show in Canton, Ohio, he would never get to play. Some people will do anything to get out of a gig in Canton, Ohio.

Now where was I before I started hearing voices in my head? Oh, yeah. It was Nashville in the early seventies. Most of the songs sounded alike, most of the singers looked alike, and most of the songwriters thought alike if they thought at all. Sound familiar? Well, that was the problem for one songwriter and pig farmer named Willie Nelson who left Music City for Texas in a daring journey some modern biblical scholars now refer to as the Exodus. He wanted to make his own music his own way and not be a slave to the record company or the powers that be. Willie was soon to lead a band of long-haired hippie cowboys farther into musical history than anyone imagined. Today he modestly says: "I just found a parade and jumped in front of it."

Waylon Jennings at the same time was fighting the same battle in Nashville. Like all of us, he struggled with his own demons as he struggled against the musical establishment. One of my first memories of Waylon was one day as I

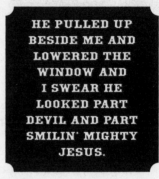

HE PULLED UP BESIDE ME AND LOWERED THE WINDOW AND I SWEAR HE LOOKED PART DEVIL AND PART SMILIN' MIGHTY JESUS.

was walking up an alley behind Music Row, and he drove up in a big Cadillac and a cloud of dust. He pulled up beside me and lowered the window and I swear he looked part devil and part smilin' mighty Jesus. On that day he gave me some words to live by that I have never forgotten. He said: "Get in, Kink. Walkin's bad for your image."

Tompall Glaser of the Glaser Brothers was the first successful Nashville cat to open up his studio to many

of us with weird songs, ideas, and hours. That was where I first met Captain Midnite, the most-often-fired disc jockey in Nashville, and a man whom, I believe, was one of the major spiritual linchpins of the whole Outlaw Movement. Midnite once stayed up for six days, told me it felt like a week, and then gave me his most cherished possession, his cowboy hat. I wore it for a while until Tompall violently yanked it from my head during a rather intense pinball game, proceeded to wear it for a while, and then gave it to Waylon.

Soon everyone was wearing hats, swapping hats, and swapping song lyrics in a spirit that hadn't been seen since God had created Nashville. Tompall claims that that pinball moment when he grabbed my hat and put it on his head without even tilting was the moment the Outlaw Movement spiritually began. Bill Monroe and Ernest Tubb, of course, he noted respectfully, had always worn hats.

Billy Joe Shaver probably was the purest, most Che Guevara–like spirit of the whole gang. In 1973 Waylon Jennings recorded an album made up almost entirely of Billy Joe Shaver songs. It was called *Honky Tonk Heroes* and it remains the very best the times had to offer.

Wanted: The Outlaws. They're wanted, all right. Today I only listen to country music on the radio at gunpoint. It seems to me to be a virtual wasteland populated by hat acts, soundalikes, and anti-Hanks. When the Outlaws were on the loose, songs were written in blood, sung by people who'd loved and cried them, lived and died them. Some of us were crucified on crosses of vinyl. Some were stoned for their ideas; stoned for their hairy, scary, soon to be legendary lifestyles; or just plain stoned. Billy Joe Shaver wrote "Honky Tonk Heroes" and we were. Lee Clayton wrote "Ladies Love Outlaws" and they did. Willie had been wandering like a modern-day Moses in the Texas desert. Waylon had been a rebel without a clause in his recording contract to say and sing what he believed. And in Austin, Jerry Jeff Walker had just thrown his new color TV into his swimming pool. As for myself, I think I was always leaving my soul at the dry cleaners in the last town we played.

SONGS WERE WRITTEN IN BLOOD, SUNG BY PEOPLE WHO'D LOVED AND CRIED THEM, LIVED AND DIED THEM.

Did the Outlaws, as they wandered through the raw poetry of time, leave any dusty dream trails for today's country artists to follow? The answer is yes and the answer is no. The only thing we can be sure of is that today's artists may for now be on the charts, but the Outlaws will always be in our hearts.

'Scuse Me While I Whip This Out

Three Wise Men

*I*t had, apparently, been one ball-dragger of a bus ride. Thirty-eight hours from the far side of Poughkeepsie to Willie Nelson's place near Austin, Texas. I phoned Willie on the bus during the waning hours of the trip and asked him if he'd like to do an interview and photo shoot for a magazine.

"Who *you*," he said in a parody of a Sly Stone falsetto. "What you *do*?"

Willie is notoriously interview-shy and, possibly because of his Native American heritage, believes that every camera that snaps his picture takes away a little piece of his soul. He doesn't dig light meters too much either.

I wanted to do the piece for several reasons: 1) I

needed to see Willie about another project and this was a good way to kill two birds and get stoned, and 2) I thought his new album, *Across the Borderline,* was some of the best work he'd done in years. Willie finally grudgingly agreed to do the interview and photo shoot under the condition that both be done on the golf course. I could live with that, even though golf wasn't my long suit. The only two good balls I ever hit was when I stepped on the garden rake.

> THE ONLY TWO GOOD BALLS I EVER HIT WAS WHEN I STEPPED ON THE GARDEN RAKE.

It was Gary Cooper time, several days later, when the two of us got together in a small alcove-like office tucked away in the bowels of the recording studio–golf course complex that is as close to a home as a jet-set gypsy like Willie Nelson ever has. First we watched a forty-five-minute video-tape of a chess match between Willie and former Texas Longhorn football coach Darryl Royal which Willie is threatening to include in a new music video.

"We've edited it down to eleven minutes," said Willie with a mischievous smile.

Somewhere before Willie mated Coach Royal (not sexually), I congratulated him on the recent settle-ment he'd made with the IRS. With tax shelters that had been disallowed and penalties, Willie at one time had owed the government over sixteen million dollars.

'Scuse Me While I Whip This Out   **11**

Many people supposedly in the know had felt that the debt was insurmountable and that Willie might well end his days as a Joe Louis–type greeter at Billy Bob's in Fort Worth. The IRS did go so far as to lock Willie's studio, seize a number of his homes, repo his secretary's old jalopy, and remove virtually everything including souvenir belt buckles, pictures, scrapbooks, and gold records on the wall. They probably would've taken Jesus if he hadn't been nailed down.

> THEY PROBABLY WOULD'VE TAKEN JESUS IF HE HADN'T BEEN NAILED DOWN.

At this writing, almost everything's been returned, and an understanding has been reached with the IRS, and Willie's suit against his former accounting firm Price-Waterhouse is still pending, all of which could result in a real financial pleasure for the Redheaded Stranger with the one-thousand-year-old bandanna.

"Uncle Sam and I are big pals now," said Willie with an expansive (and expensive) smile. "In fact, we're so close that if anybody out there owes anything to the government, they might just as well send it to me."

Willie seems quite pleased with *Across the Border-line* and well he should be. Without reviewing the album, I can tell you that Willie's version of Paul Simon's "American Tune" is worth the price of admission. It's as good as anything he's ever done and aptly demonstrates how far you can push the boundaries of country and still be country. Peter Gabriel's "Don't Give Up" with Kate Bush and Steve Bruton's "Getting Over You," which Willie performs with Bonnie Raitt, are also big winners. The most unusual song, of course, is "Heartland," which Bob Dylan and Willie wrote by fax. If it were a movie it'd require subtitles. Nonetheless, it seems to stay in your brain long after the album is over.

When we walked out in the sunlight, several friends and fellow golfers joined us and, contrasted with them, I was struck with how small Willie was for one of such legendary stature. He looked like a magical little gingerbread man and I made some comment about it.

"I'm taller than Bob Dylan and Paul Simon," he said. "Of course, they don't wear cowboy boots. I'm also taller than Ross Perot as long as I don't cut my hair."

I mentioned to Willie that a lot of superstars are short, as were Hitler, Napoleon, and Alexander the Great.

"What did they record?" he asked.

"Really tall guys almost never seem to make it," I continued, as Willie got his golf clubs. "The only two I can think of are Mick Jagger and Mick Fleetwood."

"Or anybody named Mick," he said.

I asked him if he'd noticed that Kris Kristofferson, who's also on the new album, appeared to be shrinking.

"Kris does seem to be disappearing," he said. "I've been worried about that."

That was pretty much the long and the short of it as we got into the golf cart with Willie at the wheel and zimmed off at about ninety miles an hour toward the first tee. We practically flew over a small grassy knoll at an extremely precarious angle, but the cart somehow managed to maintain its balance.

"Fortunately, we're not in control," said Willie Nelson.

Other than making music, and possibly love, playing golf is about as close to a Zen experience as Willie is likely to engage in. I asked him if he had any good golf stories he'd care to share with me.

"Well, there was this woman," he said, "who walked off this course recently complaining she'd been stung by a bee. The golf pro asked her, 'Where'd it sting you?' And the woman said, 'Between the first and second hole.' 'Well, I can tell you right now,' said the golf pro, 'your stance is too wide.'"

When Willie's not playing golf, he's mystical almost to the point of autism, which is not particularly helpful if you're trying to interview him. If you want to talk with him about ships and shoes and ceiling wax, the golf course is the place. I was determined to make the most of it.

WHEN WILLIE'S NOT PLAYING GOLF, HE'S MYSTICAL ALMOST TO THE POINT OF AUTISM.

"How does it make you feel," I said, "to know that Garth Brooks may well have sold more records this year than you and Hank Williams taken together—in your entire careers? It's a shitty question. Pretend I'm Barbara Walters."

"That is a shitty question, Barbara. I probably feel about the same way Hank does about it, which, of course, is nothing at all. Actually, I think Garth Brooks and some of the other newer artists have done a lot of good by turning millions of young people on to country music for the first time. They buy a Garth Brooks record and then they want to sample other country artists. I'm seeing a lot of younger people at my shows lately. Of course, I also saw a woman in her eighties sitting in the front row in Branson. She slept through the whole show including our rousing finale of 'Under the Double Eagle.' I appeal to all ages."

For the next hour or two, golf carts rapidly traversed the Pedernales Country Club, with Willie hitting the ball every now and then and others joining the game and dropping off in random fashion. I rode around with Willie, jotting occasional notes and working on my tan. At one point, a guy I didn't know took a big powerful swing and almost missed the ball completely. The ball traveled about eight feet at roughly a forty-five-degree angle from the tee, dribbling off like a wayward spermatozoon.

I laughed.

The guy gave me a nasty look.

> "WE ONLY HAVE ONE RULE HERE AND, UNFORTUNATELY, YOU JUST BROKE IT."

"We only have one rule here and, unfortunately, you just broke it. Never laugh at another golfer until you see him laugh first. If he doesn't laugh, you don't laugh. If he laughs, then you laugh."

"Sorry," I said. "I was pretending to be Barbara Walters."

"Have you thought of pretending to be somebody else for a while," said Nelson. "How about Morgan Fairchild?"

I'd wondered for some time how Willie had gotten involved with Paul Simon, whom I'd met in March of '92 at Farm Aid in Dallas. I asked Willie how it came about that he and Paul had worked together on *Across the Borderline*.

"When I think of me and Paul," said Willie, "it's usually Paul English I think of." English and Willie had met on the gangplank of Noah's ark and Paul, who vaguely resembles the devil and prides himself on that notion, has played drums for Nelson ever since. His appearance evokes an image strikingly dissimilar to Paul Simon's likable, middle-aged choirboy look.

"Anyway, me and Paul—Simon that is—got together like this. One day out of the blue he calls and tells me he's got a song that's 'just perfect' for me to record. I told him to go ahead and send me a tape. So he does. The song was 'Graceland.' I listened to it and called him back and told him it was a great song, of course, but it just wasn't right for me. I thanked him and let it drop at that.

"A few months later he calls and says he's leaving for his African tour but he wants to arrange studio time for me to record 'Graceland.' I tell him again it's a wonderful song and all but it just isn't right for my style. In all good faith I can't record it. I thanked him for thinking of me and forgot the whole deal.

"I didn't think anything else about it until four months later when I get a message from Paul saying that the studio time and musicians have all been set up in L.A. for me to record 'Graceland.' I figured what the hell, if he's that determined maybe he knows something I don't know. So I recorded 'Graceland.'"

"He's got pawnshop balls," I said admiringly.

"Yeah," said Willie, "but I'm taller."

As the golf balls and the afternoon rolled on we encountered Doug Holloway, who runs Pedernales Films, Nelson's movie operation. Holloway, a not unattractive young man with a fashionable ponytail, soon brought up the subject of the motion picture the three of us along with the independent producer Kent Perkins will soon be shooting. It's the film version of my second mystery novel, entitled *A Case of Lone Star*. The plot deals with a serial killer who believes he's Hank Williams and persists in knocking off country singers who play the Lone Star Cafe in New York. Willie has a seminal role in the film.

"This promises to be Willie's best role since *The Man from Snowy River*," said Holloway.

"Willie wasn't in *The Man from Snowy River*," I said.

"That's what I mean," said Holloway. "He's never had a chance to play a truly pure role with an interesting character arc that doesn't become confused with his image as a country music star."

"Every movie I make," said Willie, "I think is going to be the last one for sure. Once they see this, I figure, they'll know I went to the John Wayne School of Acting and they'll never bother me again. But they always do."

"Kind of like Paul Simon," I said.

In the film version of *A Case of Lone Star* Nelson

plays the part of Cleve, the manager of the Lone Star. I get to play myself in the movie which is one of the main reasons I got involved in its production. If a Hollywood studio had made the film, the promos would probably read: "Michael J. Fox is Kinky Friedman."

> I GET TO PLAY MYSELF IN THE MOVIE WHICH IS ONE OF THE MAIN REASONS I GOT INVOLVED IN ITS PRODUCTION.

The movie also features Ruth Buzzi as Winnie Katz, the lesbian dance instructor, Dennis Hopper and Dean Stockwell as the two New York cops, *Night Court*'s Richard Moll as Rambam the P.I. and either Dom DeLuise or Bob Dylan as Ratso, the modern-day Dr. Watson of the piece. At this writing, both men have expressed their desire to play the role. James Garner, Steve Allen, and Kris Kristofferson are also in the cast.

"One of the things I'm looking forward to," Holloway said to me, "is the duet you and Willie are doing for the soundtrack: 'Cowboys Are Frequently Secretly Fond of Each Other.'" The song, written by Ned Sublette, was slipped to Willie backstage many years ago when he guested on *Saturday Night Live.*

"I've finally found a song," said Willie, "that the two of us can do justice to."

By the end of the afternoon Willie's bandanna was

pretty damp and I was brown as a berry from riding the prairie in the golf cart. There were just two more questions I wanted to ask him.

"What do you think of the state of country music today?" I said. "All these sort of generic young artists popping out of the studio and becoming superstars, sometimes without even paying their dues. It's not as if they could go on the road again. Most of them were never there the first time."

"Phases and stages," said Willie, as he navigated the golf cart off the course and toward the recording studio. "They're as real to this generation as Bob Wills, or Spade Cooley, or Tom Mix, or Lefty Frizzell."

"Or Willie Nelson?" I asked.

"My song isn't over yet," said Willie.

We rode back in silence. Then he said: "I think the new crop is good for country music as a whole. They do seem to run the course pretty quickly, though. As I believe Garth Brooks said recently: 'Where in the hell did Billy Ray Cyrus come from?'"

"I've got one more question," I said, "then I'll leave you alone."

"Good," he said.

"Who were the most unlikely, spiritually weird golf partners you've ever had?"

"Hell, they're all spiritually weird. I've played with a lot of really great players, with presidents, senators, fugitives, pilgrims, pickers, poets. . . . And I always waited to see if they laughed first.

"There was one time," he said as he got out of the golf cart, "we had a really cosmic threesome. I never told anybody about it."

"Spit it," I said.

"It was a little secluded course in the Bahamas. I shot nine holes with John Lennon

> "I SHOT NINE HOLES WITH JOHN LENNON AND JOHN BELUSHI."

and John Belushi. We were all down there separately to get away from the world and we ran into each other."

"You are kidding," I said.

"I've never been more serious in my life."

I waited to see if he laughed. But he didn't. All he did was put away his golf clubs.

"Fortunately, we're not in control," he said.

Ode to Billy Joe

If Carl Sandburg had come from Waco, his name would have been Billy Joe Shaver. Back in the late sixties, when Christ was a cowboy, I first met Billy Joe in Nashville. We were both songwriters. Today, he's arguably the finest poet and songwriter this state has ever produced.

If you doubt my opinion, you could ask Willie Nelson or wait until you get to hillbilly heaven to ask Townes Van Zandt, who are the other folks in the equation, but they might not give you a straight answer. Willie, for instance, tends to speak only in lyrics. Just last week I was with an attractive young woman, and I said to Willie, "I'm not sure who's taller, but her ass is six inches higher than mine." He responded, "My ass is higher than both of your asses." Be that as it may, you'll rarely see Willie perform without singing Billy Joe's classic "I Been to Georgia on a Fast Train," which contains the line "I'd just like to mention that my grandma's old-age pension is the reason why I'm standin' here today." Like everything else about Billy Joe, that line is the literal truth. He is an achingly honest storyteller in a world that prefers to hear something else.

> **HE BUILT A RÉSUMÉ THAT WOULD'VE MADE JACK LONDON MILDLY PETULANT.**

Thanks to his grandma's pension, Billy Joe survived grinding poverty as a child in Corsicana. "*Course* I cana!" was his motto then, but after his grandma conked, he moved to Waco, where he built a résumé that would've made Jack London mildly petulant. He worked as a cowboy, a roughneck, a cotton picker, a chicken plucker, and a millworker (he lost three fingers at that job when he was twenty-two and later wrote these lines: "Three fingers' whiskey pleasures the drinker / Movin' does more than the drinkin' for me / Willy he tells me that doers and thinkers / Say movin's the closest thing to being free").

I believe that every culture gets what it deserves.

Ours deserves Rush Limbaugh and Dr. Laura and Garth Brooks (whom I like to refer to as the anti-Hank). But when the meaningless mainstream is forgotten, people will still remember those who struggled with success:

EVERY CULTURE GETS WHAT IT DESERVES.

van Gogh and Mozart, who were buried in paupers' graves; Hank, who died in the back of a Cadillac; and Anne Frank, who had no grave at all. I think there may be room in that shining motel of immortality for Billy Joe's timeless works, beautiful beyond words and music, written by a gypsy guitarist with three fingers missing.

Last February Billy Joe and I teamed up again to play a series of shows with Little Jewford, Jesse "Guitar" Taylor, "Sweet" Mary Hattersley, and my Lebanese friend Jimmie "Ratso" Silman. (Ratso and I have long considered ourselves to be the last true hope for peace in the Middle East.) Pieces were missing, however. God had sent a hat trick of grief to Billy Joe in a year that even Job would have thrown back. His mother, Victory, and his beloved wife, Brenda, stepped on a rainbow, and on New Year's Eve 2000, his son, Eddy, a sweet and talented guitarist, joined them. Hank and Townes also had been bugled to Jesus in the cosmic window of the New Year.

I watched Billy Joe playing with pain, the big

man engendering, perhaps not so strangely, an almost Judy Garland–like rapport with the audience. He played "Ol' Five and Dimers Like Me" (which Dylan recorded), "You Asked Me To" (which Elvis recorded),

and "Honky Tonk Heroes" (which Waylon recorded). He also played one of my favorites, which, well, Billy Joe recorded: "Our freckled faces sparkled then like diamonds in the rough / With smiles that smelled of snaggled teeth and good ol' Garrett snuff. / If I could I would be tradin' all this fatback for the lean / When Jesus was our savior and cotton was our king."

Seeing Billy Joe perform that night reminded me of a benefit we'd played in Kerrville several years before. Friends had asked me to help them save the old Arcadia Theater, and I called upon Billy Joe. Toward the end of his set, however, a rather uncomfortable moment occurred when he told the crowd, "There's one man I'd like to thank at this time." I, of course, began making my way to the stage. "That man is the reason I'm here tonight," he said.

I confidently walked in front of the whole crowd, preparing to leap onstage when he mentioned my name. "That man," said Billy Joe, "is Jesus Christ."

Much chagrined, I walked back to my seat as the audience aimed their laughter at me like the Taliban militia shooting down a Buddha. It was quite a social embarrassment for the Kinkster. But I'll get over it.

So will Billy Joe.

My Scrotum Flew Tourist: A Personal Odyssey

Forming a country-western band and calling it the Texas Jewboys was either a very smart or a very stupid thing to do. I was a Peace Corps volunteer in Borneo. I was stranded in the jungle for a year and a half once and the idea just crossed my desk. I was living in a Kayan longhouse upriver from the town of Long Lama in Sarawak. The Kayans had been headhunters as recently as World War II and they still kept souvenir skulls in hanging baskets on the porch. The skulls in baskets were to the Kayans what green hanging plants are to many nonsmoking vegetarian Rollerbladers today.

Most Americans are too civilized to hang skulls

from baskets, having been headhunters, of course, only as recently as Vietnam.

I remember we were returning from a fishing expedition one night, paddling upriver by torchlight. We were chewing betel nut and drinking tuak, a brutal, gnarly, viciously hallucinogenic wine carefully culled from the vineyards of Lord Jim.

WE WERE CHEWING BETEL NUT AND DRINKING TUAK. A BRUTAL. GNARLY. VICIOUSLY HALLUCINOGENIC WINE.

The Kayans don't give a flying Canadian whether they catch any fish or not. They claim to just be "visiting the fish." This quaint and primitively poetic little notion, unfortunately for them, does not culturally compute.

Yet I came to share their timeless, tribal outlook. I visited the fish. I watched the river flow. I got so high that I started to get lonely. It was a strange, gentle feeling, like warming your hands in a Neanderthal campfire. Not cosmic. Not mystical. But not the kind of thing you'd really want to share with the Charlie Daniels Band.

I GOT SO HIGH THAT I STARTED TO GET LONELY.

I never saw God in the jungles of Borneo, but it was during this time, on a dark, primeval night, that I did see a nine-hundred-foot Jack Ruby.

I still vividly remember what Jack said to me. He

said, "Kinky, this is Jack. I, like yourself, am a bastard child of twin cultures. You know, I just never could forgive Dallas for—what they did to Kennedy. Didn't like what they did to the Redskins either—Kinkster, baby, it's up to you now, sweetheart."

In the monsoon months ahead I became almost obsessed with Jack's messianic words. Again and again I saw him in my dreams, jumping out of the shadows.

> ON A DARK,
> PRIMEVAL NIGHT,
> I SAW A NINE-
> HUNDRED-FOOT
> JACK RUBY.

I felt his warm, comforting, sleazy presence rushing through my veins in the middle of the dank jungle night like the screaming of an endless subway circus train. I saw American dreams going up like little puffs of smoke from the infamous Texas Cookbook Suppository Building in Dallas. I was proud to share Jack Ruby's heritage. Proud to be a Texan like Jack. Proud to be a Jew like Jack. I felt almost elated that he had shot Lee Harvey Oswald. It seemed fitting and proper that one of my countrymen had taken the law into his own hands and actually assassinated the assassin. "Jesus," I remember thinking at the time. "Ol' Jack must have really had some balls!"

Years later, of course, I was a little surprised and a bit disheartened when they finally exhumed Lee Harvey Oswald's grave and found Ernest Tubb.

Jack Ruby's spirit was already abroad in that land.

I had determined to form a country music band as soon as I returned to the States, and I had sworn to myself that it would be known as Kinky Friedman and the Texas Jewboys. The torch had been passed.

A Peace Corps psychiatrist was flown in by helicopter, but by this time I was pretty much cookin' on another planet. (The only other visitors I'd had in almost twenty-four months had been my parents, Dr. and Mrs. S. Thomas Friedman from Austin, Texas, who had taken a Borneo taxi, incredibly enough, all the way to the last outpost on the river. I was, naturally, thrilled to see them. I was also rather amazed to see that the driver of the taxi was Harry Chapin.)

I WAS PRETTY MUCH COOKIN' ON ANOTHER PLANET.

The Peace Corps psychiatrist listened to a few of my songs and determined that I was definitely out where the buses don't run. Finally, much to my chagrin, the Peace Corps director ordered that I be returned immediately to my own culture. Little did he dream that what was the Peace Corps's loss was soon to become country music's loss.

I left Borneo with nothing but my guitar and my wheelbarrow. I had run into a bit of elephantiasis in the jungle and I had to carry my scrotum in a wheelbarrow.

The very next day I was winging my way back to

the States. The Peace Corps was gracious enough to buy me a first-class ticket. My scrotum flew tourist.

> **I LEFT BORNEO WITH NOTHING BUT MY GUITAR AND MY WHEELBARROW.**

I got to New York just as Robert Young began filming the first of his Sanka coffee commercials for television. These, I felt, were a step down from *Father Knows Best* but certainly a step up from *Marcus Welby, M.D.* Robert Young was, fortunately, a rather distant friend of the family. I had always admired him, and now I thought I'd drop by the studios and have a few words with the wise old bird.

When Robert saw me he was shocked and disturbed at how pale and thin I was. I weighed about twenty-nine pounds and was in a rather deep state of culture shock at the time. I told him I liked Borneo but that my Peace Corps director had recommended that I be returned to my own culture because I was getting very nervous in the service. Robert Young recommended that I and a rather irritable young airline stewardess, who was also on the set, switch to Sanka brand.

Three weeks later, Robert said, "Well, Kinky, *now* how's our returned Peace Corps volunteer feeling?" By then, I weighed about seventeen pounds and was in a severe case of culture shock.

"I'm feeling great, Robert," I said. "That god-damn Sanka brand really did the trick! In fact, I'm leaving for Texas today. You might check on that stewardess, though, if you get a chance."

The young stewardess was hanging from a shower rod right there in the studio. Robert Young walked right up to her and put his hand on her shoulder. As I walked out he smiled and I heard three short, rather hollow laughs: "Ha-ha-ha." "Maybe that's the way Robert Young always laughed," I remember thinking. But it gave me kind of a strange, gentle feeling. Kind of like warming my hands in a Neanderthal campfire.

THE SONGS I HAD WRITTEN WHILE IN BORNEO INCLUDED "RIDE 'EM JEWBOY," "WE RESERVE THE RIGHT TO REFUSE SERVICE TO YOU," AND "THEY AIN'T MAKIN' JEWS LIKE JESUS ANYMORE."

I went back to the ranch in Kerrville, Texas, to round up the band and rehearse and hit the road to country music's hall of fame (or shame, depending on how you looked at it). The songs I had written while in Borneo, including "Ride 'Em Jewboy," "We Reserve the Right to Refuse Service to You," and "They Ain't Makin' Jews Like Jesus Anymore," had a little something to offend almost everyone. I knew if I could reach just one person out there that I'd be a success. But little did I dream that I would go on to become probably the best nationally known Jewish entertainer from Texas. That is, of course, unless you want to count Tom Landry.

In those early days I could sing, burp, tell jokes, smoke a cigar, and play two instruments—the guitar and the Jewish cornet (sometimes referred to as the nose). But not unlike the great Hank Williams, I had serious problems with my personal life. It was not a pleasant sight for many audiences or fellow band members to see me wheeling my scrotum off the stage after the show into the waiting U-Haul trailer. But the band played on.

WHEN THE JEWBOYS WERE HOT THEY COULD REALLY SEND YOUR PENIS TO VENUS.

We had rehearsed for six days back at the ranch, and on the seventh day we had a sound check. The band contained many former greats and many future greats and no bass players from Los Angeles.

When the Jewboys were hot they could really send your penis to Venus. But some people and some places were not quite ready for our music. So we barrel-assed across the country—a dusty station wagon pulling a U-Haul trailer down those lost highways. From Kerrville to Nashville, from Austin to Boston, from Luckenbach to Los Angeles. Schizophrenic Sons of the

'Scuse Me While I Whip This Out

Pioneers—providing bad taste in perfect harmony—setting out to prove that the world wasn't really square.

At first, we got run out of town so often that once we didn't get to go home, take a shower, and get changed for three months. But that didn't bother us. Even our harshest critics had to admit: "Their music may occasionally suck bog water, but this band consistently smells bad." Actually, we kind of dug it. We figured we probably just smelled like real outlaws, like people smell who live in Europe.

> **WE FIGURED WE PROBABLY JUST SMELLED LIKE REAL OUTLAWS, LIKE PEOPLE SMELL WHO LIVE IN EUROPE.**

Probably the whole thing started with Bob Dylan back in Greenwich Village where he never bathed, shaved, or brushed his teeth for years at a time. The only time he ever brushed his hair was before he went to bed. I once asked Bob why he did it. He said, "You know, Kink, I gotta make a good impression on my pillow."

Pretty soon Bob had the whole country looking and smelling like Sirhan Sirhan. "Talent's one part inspiration and nine parts perspiration," Bob wrote in one of his songs. "Now, Annette Funicello, won't you lay across my big brass bed?"

One might say that Bob's total disregard for personal hygiene, either dental or mental, ended the golden age of blond-haired Aryan dominance and brought about a new kinky-headed, more funky, fairly tedious era. It marked the end for Tab Hunter, Sandra Dee, and Fabian, but it would herald a new beginning for Isaac Hayes, Ira Hayes, Woody Hayes, and Gabby Hayes. And purple haze, for that matter.

We played one of our very first gigs in Luckenbach, Texas—a small German ghost town where they still tied their shoes with little Nazis. This was before Willie or Waylon had ever heard of Luckenbach. It wasn't on the maps or the charts. The jukebox contained mostly old German drinking songs and warped Wagnerian polkas. The only two popular titles I recognized were "You Light Up My Wife" and the great all-time standard, "Send in the Kleins."

WE WERE ATTACKED BY WILD INDIANS ONSTAGE IN SAN FRANCISCO.

I was a bit nervous until I looked out over the krauts. They were big and friendly and goose-stepping in time to the music. Soon they stopped polishing their Lugers altogether, clicked their heels, and broke into a moderately Teutonic variant of the bunny hop.

The days ahead were filled with excitement for me and the Texas Jewboys. We were attacked by wild Indians onstage in San Francisco for wearing those funny little dime-store Indian war bonnets and singing a funny little Indian song, "We Are the Red Men Tall and Quaint." We were attacked by dykes on bikes in

Buffalo for singing "Get Your Biscuits in the Oven and Your Buns in the Bed." One called me a "male show business pig." We needed a police escort to get out of town. Negroes chased us in Denver. Rednecks ran us out of Nacogdoches, Texas, on two different occasions. Mild-mannered, pointy-headed, liberal Jews called us a *shande* in New York and born-again nerds in the Richie Furay Band tried to shut us down in Atlanta when I sang "Men's Room, L.A.," a religious ballad by Buck Fowler:

MEN'S ROOM, L.A.

I saw a picture yesterday
In a men's room near L.A.
Lying on the floor beside the throne
Had I not recognized the cross
I might have failed to know the boss
I thought "Lord you look neglected and alone."
I picked it up with loving care
I wondered who had placed it there
Then I saw there was no paper on the roll
I said "Lord what would you do
If you were me and I were you
Take a chance, save your pants or your soul?"

And a voice said "Kinky, this is Jesus.
I ain't square. I got these pictures everywhere
From Florida on out to Frisco Bay
So boy, if you're hung up on the pot
Feel free to use my favorite shot."
I saw a picture yesterday
In a men's room near L.A.

Finally, I had to send the Texas Jewboys off on sabbatical for a while. "When the time is right," I vowed to myself, "I'll bring them all back and give them each two or three hundred dollars." I hope someday still to make that dream a reality, though I'm not too sure about the two or three hundred dollars.

I WAS A HIGHLY AMBULATORY, SOMEWHAT UNPLEASANT AMERICAN WITH A TERMINAL CASE OF LONE STAR BEER.

The point was people were beginning to hear my songs. The point was also, rather unfortunately, right on top of my head. People were beginning to accept me for what I was—a highly ambulatory, somewhat unpleasant American with a terminal case of Lone Star Beer and a tertiary case of syphilis that I had apparently run into somewhere in the jungles of Borneo. In his unbridled eagerness to give me and my scrotum the hook, the Peace Corps doctor had overlooked the latter.

Meanwhile, I kept traveling the American countryside playing my songs, telling my jokes, and consciously infecting toilet seats practically everywhere I went. This included (in what was to prove an unfortunate career move) Kenny Rogers's brand-new forty-foot jade toilet seat.

I still vividly remember emerging from Rogers's extremely ornate dumper into his sequined living room. The Southern California sun was ricocheting ferociously from the chandelier to the swimming pool to the tennis courts and back again into my right iris.

"You ol' storyteller, you," I said humorously. "I can understand the chandelier, the swimming pool, the tennis courts—but Kenny," I asked, shaking my head incredulously, "why in the world would you need a forty-foot jade toilet seat?"

"Well, Kink, you know," he said rather wistfully, "we never had one when I was growin' up."

But "the times they were a-changin'," as Willie Nelson sang in one of his songs. Rednecks were coming out of their mobile homes, women were coming out of the kitchen, and homosexuals and Jews were coming out of the closet.

I was coming out of a men's room in Denver, Colorado. It was one of the last stops on Bob Dylan's Rolling Thunder Revue, and back then I was as happy as the shah of Iran. I had just taken a rather large and highly gratifying Nixon . . . I had walked miles and miles of bathroom tiles . . . I was thinking of many things. Weird phrases peppered my cerebellum. "Save Soviet Jews—Win Valuable Prizes" . . . "Here I sit / Straining my pooper / Tryin' to give birth / To a Texas state trooper." I flashed on

> REDNECKS WERE COMING OUT OF THEIR MOBILE HOMES. WOMEN WERE COMING OUT OF THE KITCHEN. AND HOMOSEXUALS AND JEWS WERE COMING OUT OF THE CLOSET.

other times, other dimes, other walls, other stalls, other balls, other halls, other words, other turds, other nerds . . . young couples shopping for flavored toilet soaps in Georgetown, D.C. . . . myself teaching Frisbee to the natives of Borneo . . . some of the natives stealing the Frisbees . . . using them to make their lips big . . . setting back my Frisbee program. I saw the best minds of my generation destroyed by Holiday Inn sanitary wrappers shimmering in the night . . . truck stops . . . rubber machines before the Trojan War . . . airports and runways and young couples buying ludicrous, Freudian-flavored thought associations.

When I came to, a steaming cup of Sanka brand coffee was on a tray at my side and Robert Young was smiling down at me. An orderly was wheeling a wheelbarrow with a white sheet over it out into the hallway. "What happened?" I asked. "Where am I?"

"Take it easy now, Kinkster," said Robert Young. "You've had a bad accident and you're in the Cedars of Tedium Hospital. Apparently you were run over by a bookmobile as you were coming out of a men's room in Denver, Colorado. To save your life we had to give you a transfusion using the blood of a person of the Negro persuasion."

"That's moderately unpleasant," I said.

"Well, there's a good side of things, too," said Robert Young. "Your penis just grew twelve inches. Ha-ha-ha."

Hank's Last Ride: A Movie Treatment

ACT I

It is the undecaffeinated, polio, atomic bomb ambience of the early fifties, and HERMAN P. WILLIS hesitates on the sidewalk in front of the Menger Hotel in San Antonio, Texas. The hotel is across the street from the Alamo, and Willis looks weather-beaten and old enough to have been one of its original defenders. He has a mass of unruly gray hair and a grizzled, organ-grinder's mustache. He carries a large old cardboard suitcase. Willis seems jumpy, nervous in the service, and looks around with unmistakable paranoia as a black Cadillac slowly drifts by the front of the hotel like an urban shark. Willis turns his back to the Cadillac and feigns interest in AN OLD MEXICAN selling puppets on the sidewalk. The faces of the puppets are very grotesque. One of them looks exactly like Herman P. Willis. Willis is shocked to see the likeness, but the old Mexican merely responds with an evil, knowing cackle. As the black Cadillac makes another slow pass, Willis glances around nervously and quickly enters the hotel. In the lobby A SINISTER-LOOKING MAN is smoking a cigarette by the door. As Willis enters, he sees the man nod almost imperceptibly to another man across the lobby. Willis takes a deep breath and walks up to the desk with his suitcase. The desk clerk gives him the ol' fish-eye as Willis waves off the bellman and lugs the cardboard suitcase onto the elevator. As the elevator ascends, he unbuttons his ratty old coat and adjusts a gun. Then he pulls out a pint and takes a healthy slug.

Inside the room, Willis double-locks the door, and obviously in great pain, lifts the old cardboard trunk onto the bed. He opens it and extracts two large bottles of Jim Beam, a cowboy hat, and an old guitar. Walking to the mirror he removes his gray Harpo fright wig, peels off the mustache, and stoically takes in the reflection in the glass. The man he sees is HANK WILLIAMS. Twenty-eight years old, yet already a veteran soul. At the very top of the country music world, yet gaunt, haunted, tormented, and somehow spiritually even older than Herman P. Willis. He puts on the hat, takes the guitar, goes over, and sits on a chair. Hank Williams sings "I'm So Lonesome I Could Cry."

Two guys in aluminum suits walk into a Dallas nightclub looking for its owner, *JACK RUBY*. They find him in his office speaking in Yiddish to the framed, autographed picture of Hank Williams on his desk. The mob, in an effort to legitimize itself, wants to move into the area of country music publishing. Ruby, as a minor mob character, has been "chosen" to deliver Hank's publishing to one of the mob's new companies. Ruby gazes over pro-

"IF THAT HAPPENS." SAYS ONE OF THEM TO RUBY. "WE MAY WIND UP BREAKING A LOT MORE THAN HIS HEART."

tectively at the picture of Hank and tells the hard guys that this might not be possible. They tell Ruby if he doesn't deliver fast they'll go to Hank and get the publishing. "If that happens," says one of them to Ruby, "we may wind up breaking a lot more than his heart."

HANK's in his hotel room in bed with his new wife, *BILLIE JEAN*. She's so young she looks like a kid with a big diamond on her left hand. She adores, almost venerates Hank in the near poignant fashion of an extremely soulful groupie. Hank looks tired, old beyond his years, and distracted. Part of him will always be in love with his first wife, *AUDREY*. But he loves Billie Jean as only a broken-hearted hillbilly can. Hank's back is hurting him again so he talks Billie Jean into getting on top. She accomplishes this with the practiced style of a rodeo cowgirl mounting up

> ## "SUGAR, I'VE BEEN ON TOP SO LONG I'M STARTIN' TO GET LONELY."

for what could be a rough ride. As they make love she tells him, almost innocently, that he's the first man she's met who likes for the girl to be on top. Hank looks at Billie Jean with an expression that is next door to wistful. "Sugar," he tells her, "I've been on top so long I'm startin' to get lonely."

That night Hank and his band The Drifting Cowboys perform at The Ol' Barn, a sawdust-floored honky-tonk owned by Hank's friend and fellow country star, CHARLIE WALKER. It is September 17, 1952, Hank's last birthday, and he's livin' it up like he knows it. Charlie tries to cool out Hank's drinking before the show, but Hank tells him: "Charlie, you gotta find what you like and let it kill you." Charlie shakes his head. Hank takes the stage to tumultuous cheers. He takes a swig from a bottle he's "hid" onstage, and winks at JERRY RIVERS, his fiddle player. "A little firewater for ol' Hank," he says. Hank launches into "Kawlija," a song with a heavy Indian-style tom-tom backbeat and poignant lyrics about a brokenhearted cigar store Indian who "wishes he was still an ol' pine tree." Hank drinks more and starts enjoying himself, and so does the crowd. Toward the end of the show, a strange transformation comes over Hank. He's now serious, intense, almost somber. He reintro-

Kinky Friedman

duces himself to the audience as LUKE THE DRIFTER, his professional alter ego, and performs "Lost Highway." His tone now is moralistic, almost religious in fervor. Some in the crowd weep. Then suddenly he's back to Hank, and closes the show with a cranked-up verse or two of "Jambalaya," the number-one song in the country. As Hank sings, he sees AN UNFRIENDLY LOOKING MAN who seems out of place with the rest of the crowd.

"CHARLIE, YOU GOTTA FIND WHAT YOU LIKE AND LET IT KILL YOU."

The man is definitely not into the music. He's the same man who was in the lobby of the Menger Hotel.

Later, with Billie Jean sitting shotgun at a table to keep away the country-cute barmaids, Hank slumps in a chair as CHARLIE WALKER brings out a big birthday cake, and the band, a few friends, and hangers-on gather 'round and sing a rather tepid version of "Happy Birthday." Hank is staring strangely into space, and looks more like the honoree at a wake. When he blows out the candles, a kibitzer comes over and asks him what he wished for. Hank closes his eyes. "Wishin' I was still an ol' pine tree," he says.

At 3:20 in the morning, Hank and Jim Beam call Jack Ruby from Hank's hotel room. Billie Jean tries with zero luck to get Hank to come to bed. Hank's convinced that someone's out to kill him, and relates a lurid string of attempts on his life to Ruby. Ruby's response is to ask Hank about his publishing. Hank becomes drunkenly indignant. Someone's trying to kill his ass, and Ruby's talking publishing. No, he won't give up his publishing. Ruby doesn't want Hank to know his involvement with the mob, though, of course, everyone who knows Ruby already does. Ruby lays off the talk of the publishing and reacts to the murder attempts. He has friends, and he'll take care of it. When Hank hangs up, Ruby pulls out a gun and waves it around his office in impotent rage. A janitor is sweeping the floor in the hallway and says, "Put dat thing away, Mr. Jack. One of these days you gonna shoot somebody!"

Hank decides he needs some air. Ignoring Billie Jean's pleading, he leaves the room, crosses the empty lobby, and walks out into the deserted street. He takes out a pill or two, and washes them down with a hit from his ever-present pint. He's gazing spellbound at the moonlight shining on the Alamo when a black Cadillac roars down the street directly toward him. Hank dives to get out of the headlights, and collides heavily with an old stone wall beside the historic mission. A MAN IN A BUGS BUNNY MASK leans out the window as the car passes by. Bugs fires five shots at Hank. Hank collapses in the gutter as the Cadillac tears off into the night.

Kinky Friedman

ACT II

PHILLIP SLADE, *the famous detective now in his waning years, is emptying out a cobwebbed, battle-scarred desk in his San Francisco office. His longtime secretary comes in, consoles him on the recent death of his wife, tells him the world of crime fighting won't be the same without him. When she leaves, Slade takes out a bottle of Jim Beam and pours out a stiff shot. The secretary's voice crackles over the intercom: "Mr. Jack Ruby is calling from Dallas." Ruby and Slade are apparently old friends. The nightclub owner wants Slade to find out who's been trying to kill his friend, the country star named Hank Williams. Three attempts have been made in the past few weeks. "Never heard of him," says Slade. He doesn't like to travel much these days, and country music gives him projectile vomit. Ruby's an old friend, however, and tries to call in the chip. Ruby doesn't know about Slade's getting close to the end of the line. Slade's been "between cases" for a long time. Ruby still wants him. Ruby does not tell Slade who he suspects the would-be killers are. He just asks him to protect Hank, and warns him that he may run into some tough characters. With a thin smile, Slade thanks Ruby for the tip. In giving him the case, Ruby's actually doing him a favor; giving him a new lease on life. Slade kills another shot of Jim Beam, grabs his hat, and goes home to pack.*

That afternoon Slade flies to Texas. One senses that, like an old inveterate smuggler, Slade feels this will be his last big run. He checks into a room at the Menger Hotel just down the hall from Hank and the band. Billie Jean has gone home to her family for a few days. The honeymoon is definitely over.

Hank's in bed in his hotel room having a nightmare. His chauffeur is driving him in a black Cadillac and suddenly turns off at a signpost that reads: LOST HIGHWAY. Hank tries to stop him, but when the driver turns around there is only a hideous death's-head under the chauffeur's cap. Hanks sits bolt upright in bed. He's wide awake. The rhythmic sound of the turning ceiling fan becomes the drumbeat of the song "Kawlija." Suddenly AN INDIAN wearing war paint and clutching a long knife leaps on Hank as he struggles desperately for his life. Hank is cut in several places, and the Indian escapes out the window. Slade and some of the band hear the commotion, rush in to find Hank bleeding all over the place. Slade identifies himself, takes charge, investigates the scene, looks out the third-story window with no fire escape, and shakes his head.

Hank and the band go on the road. Slade stays with them so closely that when Hank eats watermelon, Slade spits out the seeds. Hank's starting to cook on another planet.

Kinky Friedman

He's drinking heavily and taking chloral hydrate tablets some "sawbones" prescribed. In a roadside restaurant, he sees a picture on the wall of the battleship Missouri. He takes out his gun, and to the horror of the patrons,

SLADE STAYS WITH THEM SO CLOSELY THAT WHEN HANK EATS WATERMELON, SLADE SPITS OUT THE SEEDS.

shoots the picture five time. As Slade hustles him out of there, Hank shouts: "It drew on me first."

Slade reports in to Ruby. Ruby tells him to keep doing so. Hank's like a son to him. As more incidents occur and Slade fails to apprehend the criminal, Ruby becomes increasingly agitated. Slade privately wonders if he's losing his powers. Slade feels you're only as good as your last case; and the way business has been going for him lately, this could well be it. He's too old for this kind of thing, and he knows it. But somehow his whole manhood, his very existence, seems to have become inextricably bound up in this baffling case. Billie Jean comes back, and a certain rapport is established between her and Slade. Each sees the other as a possible way to reach Hank. Unfortunately as far as Hank is concerned, the lines are down. Hank's locked in mortal combat in some mysterious internal casino. Nonetheless, Hank becomes jealous of Slade and Billie Jean. Something's going on between them that he doesn't understand.

Hank becomes increasingly skeptical of Slade's abilities. His attitude becomes one of mocking the old gumshoe's efforts. Threats to his life notwithstanding, Hank's making a game of it. Slade calls Ruby to report his general frustration with the case and his belief that Hank is currently residing on the third ring of Saturn. Ruby looks at the autographed picture of Hank on his desk. He takes out his gun and says something in Yiddish.

HANK IS CURRENTLY RESIDING ON THE THIRD RING OF SATURN.

Hank and Slade are fated foils. Each man is stubborn, strong-willed, and lonely as hell, and peculiarly moral. As well as their overt animosity, a grudging, almost unconscious admiration is developing between the two. They are very much like each other, though they don't see it that way. They are worthy adversaries. At the Skyline Club in Austin, Luke the Drifter is performing the gospel tune "I Saw the Light." We keep the music, and go to Slade in a hotel hallway outside the door of Hank's room. He lets himself in with a skeleton key as Hank sings the line "Then Jesus came like a stranger in the night. Praise the Lord, I saw the light." As Hanks sings the chorus, Slade snaps on a flashlight, goes to Hank's trunk, opens it with another key and looks inside. At the bottom of the trunk, under some cowboy shirts, is a blood-covered knife.

After the show, Hank is so drunk he sits in the parking lot supporting himself against a post, still demanding to drive his Cadillac. BRACK, the road manager, forcibly grabs Hank's keys and goes to bring the car around to Hank. Hank tries to stop him, and then watches in horror as Brack starts the ignition. As he does, the car explodes into about six million pieces.

ACT III

Hank goes back to Alabama to rest before his next tour. He spends a few weeks at a sanatorium trying to dry out and get his health back, then moves to his mother LILLY's boardinghouse in Montgomery. Lilly is a tough, domineering woman who thinks nothing of getting into fistfights with her son, and who despises Billie Jean when she moves into the boardinghouse with Hank. Hank doesn't see Slade during this period, and thinks he may have shaken him. But Slade is staying in touch with Billie Jean, and also has been to see Hank's first wife, AUDREY, in Nashville. A picture is coming together in Slade's mind, and it's not something he'd want to hang in his sitting room. Since Hank's number one in the country, he feels compelled to perform, to strike while the iron's hot, even though he's obviously in no shape for the road or anywhere else. The pressures on Hank are enormous. He travels like a

gypsy, and works like a dog to support his new wife, his ex-wife and son, his mother, the band, his reputation, and all the parasites and excess baggage that come with success. Some of the things that make a country star shine are the very things that can make the star fall.

Hank does a radio ad for a station in Ohio. He closes with a favorite line of his: "If the Good Lord's willin', and the creek don't rise, I'll see you at Canton Memorial Auditorium on New Year's Day, 1953."

On Monday, December 29, the night before Hank is to leave for his tour, he and Billie Jean are talking in bed in Lilly's old boardinghouse. Billie Jean is telling Hank how as a little girl she used to talk to the old and broken Christmas tree ornaments and tell them that the back of the tree was a very important place to be. Hank listens and smiles a sad smile. He tells her that every time he closes his eyes he sees God coming down the road for him. Jesus has told him that he's going to die soon. Billie Jean tells him not to talk that way. Sitting up in bed, she rubs his back. Then Hank lays her down and gently makes love to her. This time he's on top, and both of them make love with an edge of sadness as if they know this will be the last time. Afterward as they lie next to each other, Hank says to Billie Jean: "If ever you need me and you cain't find me, look to the back of the tree."

"IF EVER YOU NEED ME AND YOU CAIN'T FIND ME. LOOK TO THE BACK OF THE TREE."

The next day, Tuesday, December 30, Hank, with his gun, pint, and pills, leaves in his chauffeur-driven white Cadillac for a New Year's Eve gig in Charleston and a big New Year's Day show in Canton, Ohio. As he speeds through his final night on the planet, Hank drinks, pops pills, scribbles lyrics furiously in the backseat of the Cadillac. The weather is dark and foreboding. Hank periodically begins to see a black Cadillac following him. The driver appears to be an angel. Or is it a devil?

Jack Ruby, in his office, opens a large envelope that has just come in the mail. In it, a notarized document formally assigns all of Hank's publishing to Ruby to publish as he sees fit. Ruby seems very sad about this, as if a cowboy who was going to die just offered him his saddle. "You shouldn't have done this, Hank," he says. "It's my fault."

Hank now notices much to his disbelief that Slade is following him in an old Oldsmobile. Hank now almost feels sorry for Slade as the veteran sleuth doggedly pursues his seemingly hopeless investigation. Hank grins to himself, and shakes his head at the thought of Slade's futile efforts. Still he's got to admire his tenacity. They meet at a diner and have it out. Slade senses just how hopeless it is, but he's reluctant to give up on Hank. Slade tries to talk Hank out of doing the tour. Hank tries to talk Slade into going back to wherever he crawled out from. Nothing is resolved. Before they leave, someone plays Tony Bennett's version of Hank's hit "Cold, Cold Heart." Slade, who's from San Francisco, loves the Tony Bennett version, though it doesn't appear to blow Hank's skirt up all that much. "Now there's a voice," says Slade, needling Hank. Hank doesn't respond. By this time, he's hearing voices of his own.

Hank continues on the road, Slade trying to keep up with him. Slade loses Hank several times, each time catching up with him. It is a strange and frantic chase involving Slade's car, and the real or imagined black Cadillac weaving in and out mystically and menacingly through the rain and fog. As the chase continues, there's an increased sense of urgency on Slade's part. Catching Hank has become a matter of life and death. On Slade's car radio he hears Hank's latest hit, "I'll Never Get Out of This World Alive."

Hank goes to the Knoxville airport. He gets on a plane to Charleston. Slade just misses the flight. He's lost him.

Slade books the next plane for Charleston, and is boarding when Hank's plane comes back to the airport because of mechanical problems. Slade hears the announcement, breaks out of line, waits for Hank, and starts to tail him. Hanks sees him and rolls his eyes. He can't believe this guy. There is a semi-chase through the airport. Slade at his age is no longer O.J. Simpson but, then again, neither is Hank. Hank eludes Slade, and gets back on the road. Hank has canceled the Charleston show, and now heads for the Canton, Ohio, New Year's Day gig. He's acting very erratically, to put it mildly.

Ruby flies to New Orleans to meet with THE BOSS OF THE MOB. He carries with him Hank's document turning over his publishing. He reads it on the plane and shakes his head in resignation. As he walks through the Quarter to the Boss's office, someone is playing "Jambalaya." Ruby hears it and in a spur-of-the-moment decision, rips up Hank's document and throws it to the winds. He meets the Boss and tells him he can't deliver Hank's publishing to the mob. For a moment Ruby's life hangs in the balance as the Boss stares at him with eyes that burn like two piss-holes in the snow. Ruby says: "I owe you a big one—whatever you want—later down the line." The Boss gives him a sinister semi-papal wave, indicating that the deal with the devil is done, and the meeting is over.

'Scuse Me While I Whip This Out

55

Slade finally runs Hank down at a truck stop in Tennessee. There's action along the way as Slade almost corners Hank at a rest stop only to find him gone with bloody "Luke" written on the bathroom mirror. The stakes get higher. Hank "playfully" shoots out one of Slade's tires on the slippery highway. Slade changes the tire, catches up with Hank, and encounters him for the last time in the truck stop parking lot. As Slade slips up on Hank, something strange is unfolding. Hank is "arguing" with Luke the Drifter, and it is every bit as violent as the demons that come out of one man's head can be. Slade sees this, and the macabre scene confirms his own private conclusions. Hank sees Slade. There is a certain spiritual High Noon flavor as the over-the-hill master detective faces the burned-out country star. Slade stands between Hank and the Cadillac, blocking his escape route. Hank walks toward Slade. Slade doesn't move. Hank's no longer monkeying with the detective. There's fire in his eyes. Slade looks grim and determined. "Get the hell outta my way," says Hank. "If I want a private dick, I'll zip down my fly." "You don't need a private dick," says Slade. "You need a checkup from the neck up." The two of them grapple fiercely beside the rain-swept nighttime highway. Slade, after a lengthy struggle, finally gets the upper hand of it. Both men are out of breath. They look like two human shipwrecks washed up on the same lonely shore. Slade thinks he's talked Hank into calling off the tour and seeking help, but when Slade goes to phone a hospital, Hank removes the distributor cap from Slade's car. When Slade returns, Hank is grinning. "You solved the mystery," says Hank. "But this is one killer even you cain't stop."

In an unself-consciously Casablanca-like farewell, Slade, in trench coat and fedora, stands alone in the cold, rainy, fog-shrouded parking lot and watches Hank climb into the backseat of his Cadillac for the last time. In a righteous rage, Slade jumps into the Olds and tries unsuccessfully to start her up. In bitter tribute, Slade takes out a pint and silently toasts Hank's taillights as they disappear from sight. Then he rushes to the telephone in a last desperate effort to get medical help for Hank. In the backseat of the Cadillac, Hank takes out a pint, washes down three or four pills with a swallow, starts writing lyrics furiously on a scrap of paper. He stops for a moment, shivers. Then he starts writing again.

The rain comes down much harder and the creek rises, forcing Hank's driver to turn off the highway onto a side road. By the time Slade and an ambulance arrive, the creek has effectively cut off the highway and the side road. Hank notices a flashing detour sign, and then a sign that reads:

'Scuse Me While I Whip This Out

LOST HIGHWAY. *His nightmare is coming true. Hank yells, and when the driver turns around, Hank's staring death in the face.*

An hour and a half later near Rutledge, Tennessee, patrolman SWANN KITTS stops Hank's seventeen-year-old driver, CHARLES CARR, for speeding. As he's writing out the ticket, patrolman Kitts glances in the back. Hank is sprawled across the backseat. His hand is clutching a scrap of paper. "That guy looks dead," says Kitts. "He's just sedated," says Carr. He pays the twenty-five-dollar ticket and goes on his way. Hank's body shifts slightly as the Cadillac accelerates. The scrap of paper flutters from his hand. Hank's last lyrics read: "We met, we lived and dear we loved, then comes that fatal day, the love that felt so dear fades far away. Tonight love hath one alone and lonesome, all that I could sing, I love you still and always will, but that's the poison we have to pay."

"I LOVE YOU STILL AND ALWAYS WILL. BUT THAT'S THE POISON WE HAVE TO PAY."

Billie Jean is at home with her MOMMA and DADDY when they get the phone call. Her daddy takes the call. When he hangs up the phone, there are sudden tears in his eyes as he hugs his daughter. "He's gone off one of the mountains," says Billie Jean, pulling away from her father's embrace. "Daddy, how bad's he hurt?" "Baby, he's dead," says her father. "No-oooo! He's not dead. He's just pretending."

Jack Ruby, flooded with grief, talks to himself and to God. "How could this happen? I should've known. I should've been there. Maybe I could've done something." He looks at Hank's picture. "The songs," he says, "they're safe, Hank. Now they belong to everybody." Jack Ruby takes Hank Williams's picture and holds it to his heart.

In his small office in San Francisco, Slade listens to the radio report of Hank's funeral. It's the biggest funeral the South has ever seen, the radio says. Slade walks over and turns it off, comes back to his desk, pours himself a drink, and walks over to the window. He kills the shot, swings the window open a few inches. There is music coming from a bar down the street. It's Hank Williams singing "Your Cheatin' Heart." Slade has another drink, puts on his hat, kills the light, goes back over to the window. He starts to close it, changes his mind, eases it open a little further. Slade stands at the window and looks out into the darkness. Hank's music seems to swell and fill the night.

PART 2

Presidents

Hail to the Kinkster

Along with *Bernard Baruch, Billy Graham,* and Jesse Jackson, I belong to a small, tight-knit fraternity whose members are known primarily for being a friend of presidents. Baruch died in 1965, so unless they dig him up and use him as a hand puppet, he can safely be said to be out of circulation. Graham is barely ambulatory these days, though he did speak recently at the National Cathedral. They tell me he still talks regularly with God. Jackson has not been welcomed by the current administration, partly because of his personal scandals and partly because his voice is too loud. That leaves me with the sometimes awesome, sometimes humbling burden of being the only active professional friend of presidents this country has left.

How did I attain this lofty position? How the hell should I know?

I was just minding my business one day seven years ago, promoting my latest mystery novel at a book signing in Austin at Barnes & Noble. I had just told the crowd that the reading and signing were free, but there would be a two-latte minimum. A guy came up to me and said, "Sign one for the president." I didn't think the book was really going to Bill Clinton, so I signed one of my standard inscriptions, "Yours in Christ" or "See you in hell," and forgot all about it.

Two weeks later the postmaster in Medina brought me an express envelope and said in an excited tone, "Kinky, you've got a letter from the White Horse Saloon! You know, that place in Nashville where they do all that line dancing!" I looked at the envelope. It did not say "White Horse." It said "White House." Inside the envelope was a letter from President Clinton, and at the bottom he had written, "I have now read all of your books—more please—I really need the laughs."

That was the beginning of a three-year pen-pal relationship, during which we discussed many things, from foreign affairs to more metaphysical matters. Regarding Israel, the former president wrote, "I appreciate what [your father] said about my friendship to Israel—I have to do it. Jesus was there too, you know!"

And speaking of why there are mostly white pigeons in Hawaii, land of mostly brown-skinned people, and dark-colored pigeons in New York, land of mostly white-skinned people, Bill wrote, "The white pigeons are in Hawaii and the dark pigeons are in New York because God seeks balance in all things. People seek logic and symmetry, which are different."

> "THE WHITE PIGEONS ARE IN HAWAII AND THE DARK PIGEONS ARE IN NEW YORK BECAUSE GOD SEEKS BALANCE IN ALL THINGS."

Our friendship culminated in January 1997, when the president invited my father and me to the White House. The event was a gala dinner for more than two hundred people, several of whom commented rather negatively about my wearing a black cowboy hat in the White House, but I didn't let it bother me. At first I couldn't find my name at a table setting, but when I did, I was surprised to discover that the card next to mine read THE PRESIDENT. Once I sat down, people stopped bitching about the cowboy hat. They said, "Who is that interesting man from Texas sitting next to the president?"

Sherry Lansing, then the president of Paramount Pictures, said that Bill had mentioned to her that my books would make wonderful movies. "But who," she asked, "do you see playing Kinky?" "I see Lionel

'Scuse Me While I Whip This Out

65

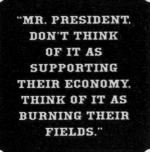

ONCE I SAT DOWN,
PEOPLE STOPPED
BITCHING
ABOUT THE
COWBOY HAT.

Richie," I said. Negotiations broke down from there. Before I left the White House that night, as a token of my gratitude, I gave Bill a Cuban cigar. I told him, "Mr. President, don't think of it as supporting their economy, think of it as burning their fields."

I first met George W. Bush about four years ago at the Texas Book Festival. At the time, he was just thinking about running for president, and I was just thinking about having another Chivas Regal. In a flash of misguided inspiration, I had taken Larry McMurtry's unclaimed name tag and slapped it on. In a

"MR. PRESIDENT,
DON'T THINK
OF IT AS
SUPPORTING
THEIR ECONOMY,
THINK OF IT AS
BURNING THEIR
FIELDS."

matter of moments people were coming up to me and telling me how much they admired my work. Not wanting to burst their bubble, and fairly hammered by then, I played along. "You've done so much for Texas, Mr. Mc-Murtry," one lady told me. "Thank you kindly," I replied. The governor, having witnessed this little exchange, eyed me quizzically.

"Look, Governor," I said. "McMurtry's a shy little booger. He'd never do this for himself. I'm just helping the old boy out with a little PR." George laughed and whispered something to several of his aides, leading me to believe I was soon to be eighty-sixed from the affair. But nothing happened. I asked one of the aides what he'd said, and he told me that the governor had said, "I want that guy for my campaign manager."

A few years later I got the first and only job I've ever had in my life, other than hand, writing for *Texas Monthly.* My maiden column, titled "All Politics Is Yokel," was about my failed race for justice of the peace in 1986 in Kerrville. Out

HE WAS JUST THINKING ABOUT RUNNING FOR PRESIDENT. AND I WAS JUST THINKING ABOUT HAVING ANOTHER CHIVAS REGAL.

of the blue I received a letter from Camp David. The folks at the Medina post office knew there wasn't any line dancing at Camp David, so they were duly impressed. The newly elected president thanked me for mentioning his name in the column and for doing so "without any curse words." He invited me to visit him and Laura at the White House, saying of Washington, "This place can sure benefit by laughing and you make us laugh." At the bottom of the letter he wrote a P.S.: "Are you running again for J.P.?"

I wrote George back and told him I had four dogs, four women, and four editors, and could he check with Laura and let me know if the four dogs and the four

'Scuse Me While I Whip This Out

> **"I DON'T KNOW ABOUT THE WOMEN," GEORGE WROTE, "BUT THE FOUR DOGS, MAYBE."**

women could sleep with me in the Lincoln Bedroom? The answer to my query came back promptly. "I don't know about the women," George wrote, "but the four dogs, maybe."

Some might ask, particularly these days, how the president can afford to maintain such a lighthearted relationship. The answer is, particularly these days, everybody needs a laugh, and everybody needs a friend.

The Houseguest

I_f you've seen one Lincoln Bedroom, to para-_
phrase my father, you've seen 'em all.

Yes, those news reports were true. I did spend a
night at the White House as a guest of President and
Mrs. Bush. It's not true that I slept in the Lincoln Bedroom, but I did visit the place, and I hung out there
long enough to bounce on the bed and soak up what
residual ambience remained after Steven Spielberg,
Barbra Streisand, and half of Hollywood had done
their best to suck out some of its soul.

Kinkster, you're probably asking, how in the hell
did you get to the White House in the first place? By
limo, of course—by special White House limo, blue

as the deep blue sea. It picked me up just in the nick of time in front of Washington's posh Willard Hotel, where I'd been waiting with my friend Jimmie "Ratso" Silman, the little Lebanese boy in my band. The doormen at the Willard were obviously ignorant of my talents, and the situation was not helped by the fact that Ratso and I were wearing large black cowboy hats and that Ratso, who looks a bit like Saddam Hussein in a jovial mood, kept shouting, "Kinky's goin' to the White House! Kinky's goin' to the White House!"

In the end, of course, Ratso was vindicated, and I was quickly whisked over to a large portico of the White House where many news photographers had gathered to watch the president depart by helicopter for a speech in Virginia. I was escorted inside and deposited at one end of a huge, ornately furnished empty corridor that wasn't empty for long, because the president, accompanied by several aides, soon appeared at the other end, rushing toward me. He gave me a big hug, then hurried for the helicopter, shouting over his shoulder, "I'll be back soon. Make yourself at home."

In a flash he was gone, and I was standing alone in the White House with my busted valise. Before long a friendly hostess came up and escorted me down the hallway. "Your room is on the third floor in the family compound," she said. "You're right across the hall from the solarium, where you can smoke your cigars. Hughie used to smoke there."

"Hughie?" I said.

"Hughie Rodham," she said. "Hillary's brother."

Some of the romance of getting to smoke a cigar in the White House is knowing that you're following in the smoke rings of great men. Maybe you're puffing peacefully away in the very chair where Thomas Jefferson once stoked a stogie. Or Teddy Roosevelt, Mark Twain, or Winston Churchill. Somehow Hughie Rodham wasn't quite the historical predecessor I was hoping for.

After dropping off my stuff, I took a stroll through the solarium and exited a side door onto the balcony. I found a chair, lit a cigar, and looked over the foreboding landscape of the nation's capital. The time and place were not lost upon me. It was December 7, 2001, Pearl Harbor Day, and the whole country was waiting for the other terrorist shoe to drop, and I was sitting on a balcony at the White House, what could well be the prime target of the enemy. I glanced up at the roof and saw two ninja-like figures, dressed entirely in black, creeping along the roof with automatic weapons. "I hope they're ours," I said to the Washington Monument. "Stand tall," the monument replied.

By the time I got back to my room, a tray of rather

coochi-poochi-boomalini hors d'oeuvres had been placed on the table and a beautiful hand-blown Christmas tree ornament in the shape of the White House and signed by President and Mrs. Bush was nestled in gift-wrapping on my bed. As a proud Red Sea pedestrian, I normally don't have a lot of uses for Christmas tree ornaments. I figured I'd either have to hang it or hang myself, and at the moment, I was leaning toward the latter. I hadn't seen a human being in quite a while now, and the somewhat disturbing notion crossed my mind that if the president didn't come back soon, I might have to become an Alexander Haig impersonator and take over the government.

> AS A PROUD RED SEA PEDESTRIAN, I NORMALLY DON'T HAVE A LOT OF USES FOR CHRISTMAS TREE ORNAMENTS.

Several hours later my fears were assuaged as I joined the president and a group of about forty family members and friends in the State Dining Room for dinner. I found myself seated at Laura Bush's right. Politically speaking, I'm not sure if I'm to the right of Laura Bush, but she's a woman who definitely has her own ideas. I took a chance and asked her if she'd headline a benefit in Austin for the Utopia Animal Rescue Ranch. I told her that I'd already asked Willie Nelson, who'd said yes and then backed out (Willie will say yes to anything that's more than two weeks away). I mentioned that I'd also approached Lyle Lovett but that was just before Lyle had been approached by a large, angry bull. "So I'm your third choice," the first lady said thoughtfully. "Okay, I'll do it." Thanks to her, the event was a big financial pleasure.

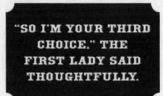

"SO I'M YOUR THIRD CHOICE," THE FIRST LADY SAID THOUGHTFULLY.

After dinner, the president asked me to read a few of my columns from *Texas Monthly,* which I did, working without a microphone, walking around between the tables like a slightly ill mariachi. It was a heady experience for a young cowboy who hadn't actively supported a political candidate since Ralph Yarborough went to Jesus. Friendship, it seems, can sometimes transcend politics. I must admit that I now feel close to the Bush family in the same way that I feel close to the Willie Nelson family and the Charles Manson family. (Maybe I'm just a lonely guy looking for a family.)

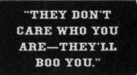

"THEY DON'T CARE WHO YOU ARE—THEY'LL BOO YOU."

What I remember most vividly about that night was smoking cigars with the president on the Truman Balcony and talking about baseball—specifically, his throwing out the first pitch at the World Series. With the excitement of being on the field at Yankee Stadium and the threat of a terrorist attack, how, I wanted to know, had he managed to toss a perfect strike? None of that bothered him, he said. What was on his mind was something that Derek Jeter, the Yankees short-stop, had told him before the game: "Whatever you do, don't bounce the ball on the way to the plate. They don't care who you are—they'll boo you."

Oaf of Office

At the White House recently, as I was advising the president on Iraq, I thought I heard him mention putting me in charge of the National Park Service. It's about time he appointed a real cowboy, I said to myself as I adjusted my fanny pack. I knew, of course, that this would not be a Cabinet-level appointment. I would probably have a portfolio of some kind, a small staff of busy little bureaucrats to do all the work, and I'd no doubt carry the title of Undersecretary. That was fine with me. I'd probably be spending most of my time under my secretary anyway.

As I left the East Wing, I became more and more

certain that, if not exactly offering me a position, the president was subtly encouraging me to get back into politics. I'd been on hiatus since my unsuccessful campaign for justice of the peace in Kerrville in 1986, but that was then and this was now. I remembered a letter he'd written me soon after he got elected, the one where he thanked me for mentioning his name in this column "without using any curse words," and asked me if I intended to run for J.P. again. On a cold day in Jerusalem, I'd

I COULDN'T DECIDE WHETHER TO KILL MYSELF OR GET A HAIRCUT.

told him. Now, after practically getting vetted by the leader of the free world himself, I found myself standing ready to serve. I waited in the full-crouch position for a call from someone in the administration. When it didn't come, I couldn't decide whether to kill myself or get a haircut.

There must be a place in politics for a man of my talents, I thought. Dogcatcher might be a good place to start. I could free all the dogs and encourage them to lay some serious cable in the yards of people I didn't like. Or maybe I could run for mayor of Austin. I could fight against the fascist anti-smoking laws. I probably couldn't win without appealing to all the techno geeks, but a good slogan would help—something like

"A cigar in every mouth and a chip on every shoulder" might work. Basically, though, it'd be too small a gig for the Kinkster. I have bigger fish to fry.

Well, what about governor of Texas? That might be therapeutic, I figured. It's a notoriously easy gig, and I'd certainly be the most colorful candidate to run for the office since Pappy O'Daniel. Hell, compared with Tony Sanchez, I'm practically Mr. Charismo. I can work a room better than anyone since the late John Tower went to that great caucus in the sky. When I meet a potential voter, I'm good for precisely three minutes of superficial charm. If I stay for five minutes, I can almost see the pity in the person's eyes.

Nevertheless, being governor might be a lot of fun. I've known the last two and a half governors, and they didn't seem to be working all that hard. Of course, I saw them mostly at social events. In fact, I once mistook Rick Perry for a wine steward at the governor's mansion. I said, "Hey, you look familiar. Do I know you?" And he said, "Yeah. I'm the governor, Rick Perry." Then he gave me a friendly handshake and looked at me like he was trying to establish eye contact with a unicorn. He didn't seem as funny as

> I'VE KNOWN THE LAST TWO AND A HALF GOVERNORS, AND THEY DIDN'T SEEM TO BE WORKING ALL THAT HARD.

Ann Richards or George W., but he came off like a pretty nice guy—for an Aggie.

After some minor soul-searching, I decided to throw my ten-gallon yarmulke in the ring and form an exploratory committee headed by the dead Dutch explorer Sir Wilhelm Rumphumper. The committee had one meeting and came back with the consensus that, as long as Willie Nelson or Pat Green didn't decide to run, I could be the next governor. They offered the opinion that many of Pat's people were probably too young to vote and that Willie, God bless him, did not quite present as clean-cut an image as I did. Also, neither of them had had any previous experience in politics. Not only had I run for J.P., but I'd also been chairman of the Gay Texans for Phil Gramm committee.

> I'D ALSO BEEN CHAIRMAN OF THE GAY TEXANS FOR PHIL GRAMM COMMITTEE.

Though the report was encouraging, I had to admit that I was beginning to find the prospect of the governorship rather limiting. I aspired to inspire before I expired. There had to be something I could do for my country besides flying five American flags from my pickup truck and telling the guy with four American flags on his pickup truck, "Go back to Afghanistan, you communist bastard!" But what of the president's offer? To clear the boards for my race for governor, I asked my old Austin High pal Billy Gammon to check the status of my appointment. Billy's so close to the president he gets to fly in the White House helicopter. "I stand ready to serve!" I told Billy.

"I'll get back to you," he said.

And he did. He told me that while the Bushes obviously considered the Kinkster a dear friend, the president had only been engaging in light banter when he'd raised the possibility of my moving into public housing. Billy also said that though I might have made a "fantastic contribution to government," George W. could see how I might well have been a "scandal waiting to happen."

> GEORGE W. COULD SEE HOW I MIGHT WELL HAVE BEEN A "SCANDAL WAITING TO HAPPEN."

"That's good news," I told Billy, "because I'm running for governor and I'd like you to be my campaign manager."

"I'll get back to you," he said.

I got over my disappointment quickly. Like most of us, I determined that I'd rather be a large part of the problem than a small part of the solution. Besides, I've got Big Mo on my side. I'm not sure how traveling with a large homosexual will go down with the voters, but hell, I'll try anything. Just this morning, for in-

> **I'D RATHER BE A LARGE PART OF THE PROBLEM THAN A SMALL PART OF THE SOLUTION.**

stance, I tried making a campaign pitch to my fellow passengers on a crowded elevator. After several of them threatened to call 911, however, it unfortunately put me a little off-message. "Now that I have you people all together," I told them, "I can't remember what I wanted you for."

Mad Cowboy Disease

In The Innocents Abroad *Mark Twain* observed, "They spell it Vinci and pronounce it Vinchy; foreigners always spell better than they pronounce." Twain didn't mention it, but they also spell better than they smell. All in all, very little seems to have changed since his time. There's nothing like a trip across the old herring pond to make you glad that you live in the good ol' U.S.A.

I knew that early March wasn't the best time to be a cowboy in Europe, yet I felt I had to honor a commitment I'd made to address an event in London with an unfortunate title: "Murder at Jewish Book Week." Everyone told me it was sheer idiocy to travel overseas with the triple threats of war, terror, and customs inspectors taking away my Cuban cigars. Yet, strangely, it wasn't courage that compelled me to go. It was simply that I was afraid at that late date to tell the lady I was canceling.

The flight was nine and a half hours long. It seemed as if almost every passenger besides myself was 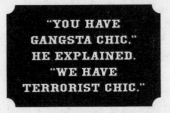 dressed in some form or other of Middle Eastern garb. One young man who spoke English was wearing a Muslim prayer cap and robe over a University of Texas sweatshirt. He told me there was really nothing to be concerned about. "You have gangsta chic," he explained. "We have terrorist chic." I found his calm analysis oddly comforting.

I was totally jet-lagged when I arrived at London's Gatwick Airport at 6:55 in the morning. My ride into town was arranged by Robert MacNeil of the old *MacNeil/Lehrer NewsHour*. (The day before, I'd been filming a PBS show with Robert in Bandera and had warned him about crossing the busy streets of the little cowboy town. "It'd seem quite ridiculous," I'd told him, "for a cosmopolitan figure like yourself to get run over in Bandera." MacNeil just said that he didn't want the headline to read "Kinky Friedman Sees Man Killed.") As I walked the cobbled streets, visited pubs and restaurants, played songs, and did interviews with the BBC, the subject of President Bush and Iraq popped up often, sometimes acrimoniously. I found myself defending my president, my country, and my cowboy hat. Soon I was going

> **I FOUND MYSELF DEFENDING MY PRESIDENT. MY COUNTRY. AND MY COWBOY HAT.**

on the preemptive attack myself, calling every mild-mannered Brit who engaged me in conversation a "crumpet-chomping Neville Chamberlain surrender monkey." After a while I realized the futility of this approach and merely told people that I was from a mental hospital and was going to kill them.

Bright and early the next morning, my journalist friend Ned Temko took me on a quest for Cuban cigars, which are legal in London, if expensive (everything is legal in London, if expensive). Phil the Tobacconist mentioned that Fidel Castro personally supplies Cuban cigars to Saddam Hussein. "I wouldn't write about that," said Ned. "George W. might nuke Fidel." As the three of us entered the walk-in humidor, Ned revealed that he'd once covered Iraq for the *Christian Science Monitor* in the late seventies. "Saddam's a thug with an excellent tailor," Ned said.

"I know his tailor," Phil said. "He's right down the street, in Savile Row."

Meeting Saddam's tailor is almost as special as meeting Gandhi's barber, but I felt I had to try. Ned,

> I TOLD PEOPLE THAT I WAS FROM A MENTAL HOSPITAL AND WAS GOING TO KILL THEM.

my Virgil of Savile Row, led us down the winding streets to a discreet-looking row of shops where tweeds were being measured for dukes and dictators behind closed shutters. Maybe it was the cowboy hat and high rodeo drag that prevented entry, or maybe it was simply the lack of an appointment, but at the designated address, no one came to the door. My outfit did get an enthusiastic response, however, from a group of city workers repairing the street nearby. They stopped what they were doing and sang cheerfully together, "I'm a rhinestone cowboy!"

"Since we didn't see Saddam's tailor," Ned said, "why don't we try to meet Tony Blair?"

"Jesus," I said.

"That's what the Yanks may think," said Ned. "Over here they're about to crucify him."

Twenty minutes later we were standing next to a Wimbledon-style grass tennis court hidden in the heart of London. "We may be in luck," said Ned. "There's Mike Levy." Levy, Ned explained, was a former record producer who'd given the world early-seventies glam rocker Alvin Stardust. "What's he done for us lately?" I wanted to know.

"He's Tony Blair's tennis partner," he said.

Levy was in a hurry, and it didn't seem likely that

Blair had played tennis that morning. Still, ever the innocent American, I stepped forward as Levy was climbing into his roadster. "Anything you'd like to say about Tony Blair?" I asked.

"Yes," Levy said. "He needs to work on his backhand."

On my last night in London, I walked through the fog until I came to the most famous address in the world, 221B Baker Street. On the door was a small bronze plaque that read "Visitors for Mr. Sherlock Holmes or Doctor Watson please ring the bell." I rang the bell, walked up one flight of seventeen steps, and suddenly I was standing in Sherlock Holmes's living room. There was a cheery fire in the fireplace. Holmes's violin stood poignantly nearby, along with the old Persian slipper where he kept his Turkish tobacco. And in the room were Japanese, Russians, Africans, people from seemingly every nation on earth, all bound together by a common, passionate belief that Sherlock Holmes was real. It was, I thought, a perfect United Nations.

The next morning I was waiting in line at the airport to board a plane back to the States. Behind me was a proper British couple with a shy little girl clutching her teddy bear and staring intently at my hat. "Ever seen a real cowboy before?" I asked.

"No," she said. "But I've seen a cow."

A Pair of Jacks

On November 22, 1963—*the fateful day that* shook the world, the day that caused Walter Cronkite to shed a tear on national television, the day that belied Nellie Connally's encouraging words, "You can't say Dallas doesn't love you, Mr. President," the day that gave Oliver Stone an idea for a screenplay—I was a freshman at the University of Texas, sleeping off a beer party from the night before. Indeed, I slept through the assassination of John F. Kennedy like a bad dream and, upon waking, retained one seemingly nonsensical phrase: "Texas Cookbook Suppository."

It was only later, once I'd sobered up, that I realized I'd been sleeping not only through history class but history itself. I'd also slept through anthropology class, where I'd received some rather caustic remarks from my red-bearded professor for a humorous monograph I'd written on the Flathead Indians of Montana. I'd gotten an A on the paper, along with the comment, "Your style has got to go." But I realized that he was wrong. Style is everything in this world. J.F.K.'s style made him who he was. Even dead, he had a lingering charisma that

STYLE IS EVERYTHING IN THIS WORLD.

caused me to join the Peace Corps. Yet, it was the style of another man in Dallas that was to change my life, I now believe, even more profoundly. I'm referring, of course, to that patriot, that hero, that villain, that famously flamboyant scoundrel, Jack Ruby.

Like the first real cowboy spotted by a child, Ruby made an indelible impression upon my youthful consciousness. He was the first Texas Jewboy I ever saw. There he stood, like a good cowboy, like a good Jew, wearing his hat indoors, shooting the bad guy

THAT PATRIOT. THAT HERO. THAT VILLAIN. THAT FAMOUSLY FLAMBOYANT SCOUNDREL. JACK RUBY.

who'd killed the president, and doing it right there on live TV. Never mind that the bad guy had yet to be indicted or convicted; never mind that he was a captive in handcuffs carefully "guarded" by the Dallas cops. Those are mere details relegated to the footnotes and

> RUBY HAD DONE
> WHAT EVERY
> GOOD GOD-
> FEARING. RED-
> BLOODED
> AMERICAN HAD
> WISHED HE
> COULD DO. AND
> HE WAS ONE OF
> OUR BOYS!

footprints of history. Ruby had done what every good God-fearing, red-blooded American had wished he could do. And he was one of our boys!

Ten years later, in 1973, with Ruby still in mind as a spiritual role model, I formed the band Kinky Friedman and the Texas Jewboys. It wouldn't have happened, I feel sure, without the influence of Jack Ruby, that bastard child of twin cultures, death-bound and desperately determined to leave his mark on the world. While many saw Ruby as a caricature or a buffoon, I saw in him the perfect blending of East and West—the Jew, forever seeking the freedom to be who he was, and the cowboy, forever craving that same metaphysical elbow room. I, perhaps naively, perceived him as a member of two lost tribes, each a vanishing breed, each blessed, cursed, and chosen to wander.

In the days and months that followed the assassination, as Ruby languished in jail, the world learned more about this vigilante visionary, this angst-ridden avenging angel. Ruby, it emerged, was indubitably an interesting customer. He owned a strip club in which the girls adored him and in which he would periodically punch out unruly patrons. This cowboy exuberance was invariably followed by Jewish guilt. Josh Alan Friedman, a guitar virtuoso who is as close to a biographer as Ruby probably has, notes that Jack was known to pay medical and dental bills for his punchout victims and offer them free patronage at his strip club. With Lee Harvey Oswald, however, this beneficence was not in evidence. According to Friedman, Ruby was utterly without remorse over Oswald's death, delighting in the bags of fan mail he received in his prison cell.

THE JEW, FOREVER SEEKING THE FREEDOM TO BE WHO HE WAS. AND THE COWBOY, FOREVER CRAVING THAT SAME METAPHYSICAL ELBOW ROOM.

In time the mail petered out and, not long after that, so did Ruby. He died a bitter man, possibly the last living piece in a puzzle only God or Agatha Christie could have created. I didn't really blame Ruby for being somewhat bitter. The way I saw it, he *had* actually accomplished something in killing Oswald. He'd helped one neurotic Jew, namely myself, come up with a pretty good name for his band.

Years after Ruby had gone to that grassy knoll in the sky, my friend Mickey Raphael, who plays blues harp with Willie Nelson, tried to get a gig at Jack's old strip club. At the time, Mickey had a jug band, and though he found the place to be redolent of Ruby's spirit, he didn't get the gig. "I thought you guys *liked* jugs," Mickey told the manager.

A PAIR OF JACKS. ONE REMEMBERED WITH THE PASSION OF AN ETERNAL FLAME. THE OTHER ALL BUT FORGOTTEN.

Thus is the legacy of one little man determined to take the law into his own little hand. And so they will go together into history, a pair of Jacks, one dealt a fatal blow in the prime of his life, the other dealt from the bottom of the deck; one remembered with the passion of an eternal flame, the other all but forgotten. Friedman notes that Ruby wept for Kennedy. Chet Flippo, in his definitive book *Your Cheatin' Heart,* tells of Ruby's friendship and loyalty a decade earlier toward another one of life's great death-bound passengers, Hank Williams. Ruby, according to Flippo, was one of the last promoters to continue to book Hank as the legend drunkenly, tragically struggled to get out of this world alive. He was also one of the few human beings on the planet who knew Hank Williams *and* spoke Yiddish.

Was Ruby a slightly weather-beaten patriotic hero? Was he a sleazeball with a heart of gold? Was he following a star, trying to find a manger in Dallas? My old pal Vaughn Meader, who in the early sixties recorded the hugely successful *The First Family* album satirizing J.F.K., probably expressed it best. After flying for most of that tragic day, oblivious to the news, he got into a taxi at the airport in Milwaukee. The driver asked him, "Did you hear about the president getting shot?" "No," said Vaughn. "How does it go?"

PART 3

Other Troublemakers

Death of a Troublemaker: A Cowboy's Elegy for Irv Rubin

To be a hero in today's world, first you have to be dead. Then you have to have been a great troublemaker, a person not afraid to stir the putrid pot of human history from time to time, causing the Starbucks patrons of the world, and who among us is not a Starbucks patron, to see our glazed, rather ill-defined reflections in our latte, and just possibly reexamine who we are and why in God's name we are here.

Such a troublemaker, I believe, was the late Irv Rubin. Certainly, Irv followed in a long line of great Jewish troublemakers, beginning with Moses and, of course, Jesus, who was undeniably a great teacher but just didn't publish. There is a long line of Jewish troublemakers who followed in their footsteps, Spinoza,

Karl Marx, Groucho Marx, Lenny Bruce, Joseph Heller, Bob Dylan, Abbie Hoffman, and John Lennon (who should've been Jewish), to name just a few, men who were, in their own ways, sentient lighthouses in a sea of humanity.

Does Irv Rubin qualify for membership in this vaunted pantheon of great Jewish troublemakers? Was he merely a misguided SuperJew? Was he a guy who said: "It's a dirty job and I get to do it?" Time will tell. Time and the input of all the rest of us Red Sea pedestrians, we who shun trouble, we who avoid at all costs confrontation, we who wait in long lines with our kids to see *Spider-Man*, we who look in our mirrors some mornings and are mildly surprised to find that we're still here.

> WE THE PEOPLE WHO SURVIVED THE CRUSADES, THE INQUISITION, AND THE HOLOCAUST JUST WANT A GOOD TABLE IN A RESTAURANT.

We the people who survived the Crusades, the Inquisition, and the Holocaust just want a good table in a restaurant. We don't want to infiltrate neo-Nazi gangs and destroy them. We don't want to spend our lives being harassed by the F.B.I. and framed incessantly by cops. We don't want to be murdered in a lonely prison cell. We want peace. We'll send money, we'll plant trees, we'll do anything short of fighting for Israel. And, if we wake up one day and Israel is no more, we will see that she is buried in a Jewish grave. And we will be very sad, indeed. We will, in fact, mourn for ourselves. For we are the children of Israel.

But far be it from me to be casting asparagus upon the American Jew. I am an American Jew and proud of it. Yet we seem to be disappearing faster than stock dividends these days. Maybe it's because we're culturally deprived. Growing up in America, we were never allowed to hear the three words this country loves best: "Attention Wal-Mart shoppers!" Even Alan Dershowitz has noticed the phenomenon of assimilation, having written about it in one of his recent books, *The Vanishing American Jew*, which some had hoped would be autobiographical. But he still had a point and I still had a point before I began hearing voices in my head. The point is, if you do not stand for something, you do not stand at all.

I'm not making light of Irv Rubin. I just think anything worth crying can be smiled. If overzealousness made aspects of his life a comedy of errors, his passion and his take on the truth made aspects of his death an American tragedy. To my mind he is a Jewish John Brown. A Joe Hill. A smilin' Jesus of the California night.

My father once said that a man can be judged by the size of his enemies. Between the windmill and the world, will we leave Mozart in the gutter, Rosa Parks

> MY FATHER ONCE SAID THAT A MAN CAN BE JUDGED BY THE SIZE OF HIS ENEMIES.

in the back of the bus, Sharansky in the gulag, Anne Frank in the attic? Yes, we will. We've done it before and we'll do it again. It's not our job to save the planet or even to save the Jews. We'll leave that to Don Quixote and Irv Rubin. For if ever there was a quixotic spirit in this troubled world, it was Irv. Shall we rage against the violent death of a sometimes violent man? Shall we keep him at arm's length even if he was on our side? Shall we bow our heads and mourn the death of SuperJew?

No need to, folks. He's taken his place in that eternal chain of great Jewish troublemakers who at most have altered the history of mankind, and at least have kept the human condition from becoming stultifyingly dull. Personally, I think our boy belongs with Moses and Jesus. Many passive, officious, humorless, and constipated people may, indeed, disagree. But even they would have to say this for Irv Rubin: "He was a fool, God bless him."

Tangled Up in Bob

I t was the fall of 1973, and my band, the Texas Jewboys, was playing the Troubadour in Los Angeles. One night Bob Dylan walked in barefoot, wearing a white robe. Possibly he thought he was Jesus Christ or Johnny Appleseed, or maybe he'd just gotten out of the bath, but everybody definitely treated him like a god. He was friendly, cryptic, and almost shy when he was introduced to us after the show.

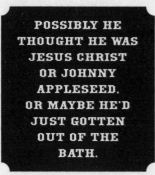

POSSIBLY HE THOUGHT HE WAS JESUS CHRIST OR JOHNNY APPLESEED, OR MAYBE HE'D JUST GOTTEN OUT OF THE BATH.

Later, we watched from the dressing room window as he got into his limo in the alley behind the club. Willie Fong Young, our bass player, said it best at the time: "He may not have any shoes, but at least he's got a limo."

It wasn't long after that that his road manager called my road manager (who, cosmically enough, was named Dylan Ferrero). I was instructed to go out on the Santa Monica pier at midnight and meet a baby-blue 1960 Cadillac convertible that would take me to Bob. After a long, mystical journey, I wound up at the home of Roger McGuinn, the founder of the Byrds, who was to become a friend of mine even though I did make the following comment to him that night: "There is a time to live and a time to die and a time to stop listening to albums by the Byrds."

By two o'clock in the morning, I had still not seen Bob, but I did stumble upon Kris Kristofferson talking to a young groupie he'd apparently just met. Kris looked up and said, "Kinky?" Simultaneously, the girl and I responded, "Yes." Kris pointed me in the direction of the kitchen. I wandered in, and there was Bob sitting on the counter, strumming a guitar and singing a song I'd written, "Ride 'Em Jewboy."

It was fashionable in the early seventies to talk long into the night about "life and life only," and Bob and I did that. I told him about my recent trip with the members of Led Zeppelin aboard the *Starship,* their private jet with a fireplace, and that I was particularly excited about urinating backstage next to Jimmy

Page. Bob was not impressed. "They have nothing to say," he said. "You and Kris have a lot to say. You should say it. Without," he went on, "using makeup and dry ice."

I WAS PARTICULARLY EXCITED ABOUT URINATING BACKSTAGE NEXT TO JIMMY PAGE.

Later, I went off to find a drink, and when I returned, Roger was helping Bob up off the floor. "The wine's not agreeing with him," Roger said. That night, I suppose, I wasn't agreeing with him much either, but that could have been because I had a chip the size of Dallas on my shoulder. Or it could just be that time changes the river. However you look at it, it's now clear that Led Zep, like so many other acts, has been relegated to the bone orchard of nostalgia, while Bob remains a spiritual beacon in a world largely remarkable for its unwillingness to be led to the light.

Traveling and making music with Bob is a rare opportunity to see a magic messenger at work and play. In 1976 Bob asked me to join him and Joan Baez, Joni Mitchell, Eric Clapton, Ringo Starr, Allen Ginsberg, and many others as part of his Rolling Thunder Revue, which traveled across America that year leaving behind some satisfied women, some wildly enthusiastic audiences, and some brain cells that promised they'd get back to us later.

I hung out a lot with Bob after that tour, and as

mesmerizing and untouchable as he seems onstage, offstage he can be extremely warm and witty. Imagine Bob and me standing in the parking lot of a seedy motel in Fort Worth at two-thirty in the morning with a redneck motel manager repeatedly asking him for his driver's license. Or picture Bob at a barbecue at my parents' house in northwest Austin. (When my mother brought him a plate, he said, "Thanks, Mrs. Friedman. You must be very proud of your son.") I remember shopping with Bob at the famous Nudie's in North Hollywood, where he saw a rhinestone jacket embroidered with Jesus's face. "A guy ordered this a long time ago," Nudie told Bob, "but he never came back for it." "He has now," said Bob. Bob bought the jacket, wore it for one performance, and then gave it to me. The Bob Dylan Jesus Jacket promptly brought me seven years of bad luck, after which I sold it at Sotheby's. (It hung for a while in the Hard Rock Cafe in Tel Aviv.) Several years ago I caught up with Bob in New York and told him what I'd done with the jacket. He shook his head and said, "Bad move."

Speaking of jackets, I once spent a month with Bob in the village of Yelapa, off the western coast of

Mexico. Although it was over one hundred degrees every day, Bob never took off his heavy leather jacket. I knew he was from Minnesota, but it did seem somewhat odd, so one day on the beach I asked him about it. His answer was to tell me a story about the king of the gypsies, and how, when the king got old, all his wives and children left him. I thought at the time that Bob might be feeling a chill that few of us ever feel.

People often ask me what Bob is really like. He's naturally shy and superstitious and hates to be photographed because he believes that every picture taken of him reduces his chances of becoming an Indian when he grows up. Bob, in fact, has a lot in common with the Native American people. They both believe, for instance, that you can't own land or a waterfall or a horse. The only thing they both believe you can own is a casino. Yet Bob's been so many things in his life that it's almost impossible to pin him down. He's been a vegetarian, an Orthodox Jew, a born-again Christian, a Buddhist, a poet, a pilgrim, a picker, a boxer, a biker, a hermit, a chess player, a beekeeper, and an adult stamp collector—and almost everything, except a Republican, that a human being can possibly be when a restless soul is forever evolving toward his childhood night-light.

And, of course, he's a very funny American. I remember once when we had to book a flight at the last minute and there was nothing in first class available. When we got back to coach, there were only a few seats left and Bob found, much to his dismay, that he was seated next to an enthusiastic young female fan. "I can't believe I'm sitting next to Bob Dylan!" she screamed. Bob gazed calmly at the girl. "Pinch yourself," he said.

The Four Horsemen
of the Antipodes

Robert Louis Stevenson once said: *"To travel hopefully is a better thing than to arrive."* This is particularly true, of course, if you've lost your luggage. In the fall of 1985, realizing that it's sometimes best to leave life's excess baggage behind, my father and I finally agreed upon something. We decided to take a trip together to Australia.

Our publicly stated purpose was to put to rest once and for all two eternally vexing questions that have troubled mankind down through the centuries: Does water swirl counterclockwise in Australian toilets? And do Australian dogs circle counterclockwise three times before flopping down in the dust? Possibly a phone call to friends in Sydney might've helped resolve these matters, but, like all hopeful travelers, we wanted to find out for ourselves.

It is never highly recommended for a father and his child to travel together upon such an extensive journey, particularly if the child is almost forty-two years old and both father and son derive from a small, ill-tempered family. If indeed this is the case, human buffers are definitely in order to keep a peaceful pilgrimage from boomeranging into a rather repellent tension convention.

When you ask people if they'd like to go to Australia, they invariably tell you it's the dream of their lives. They'd love to go, they all say, but, fortunately for Australia, most of them never do. They cling perversely to that tar baby that is the quiet desperation of their lives. If they have the time, they don't have the money. If they have the money, they can't spare the time. Few of the folks we talked to realized that time is the money of love. What they *did* realize apparently was that traveling halfway around the world with Dr. Tom Friedman and his son, Kinky Friedman, could make for an extremely unpleasant experience.

After searching the American countryside high and low, Tom and I at last came upon two brave souls who stood ready to accompany us on our father-and-son odyssey. One was Earl Buckelew, an old-time rancher and neighbor of ours who hadn't been outside of Medina, Texas, in more than fifty years. Earl was old enough to be a member of the Shalom Retirement Village People, but youthful in spirit and more vigorous than many men younger than he, Tom and myself included. When he'd gone to see his doctor for a pre-trip checkup, the "ol' sawbones" had warned him to be careful about his high cholesterol. "Hell," Earl had told the doctor, "when I was growin' up we didn't even know we had blood."

> "HELL," EARL HAD TOLD THE DOCTOR. "WHEN I WAS GROWIN' UP WE DIDN'T EVEN KNOW WE HAD BLOOD."

The other candidate for the journey was my long-time journalist friend from New York, Mike McGovern. McGovern was a large half Native American, half Irish legend who rarely had much trouble getting into trouble. He was, however, not without charm. McGovern, reportedly, had once combed his hair before meeting a racehorse. Taken together, I'd always felt that McGovern and I just about comprised one adequate human being.

The four of us taken together, however, comprised a new, rather reckless, dangerously unstable entity. McGovern, ever the adept headline writer, dubbed our little troupe the Four Horsemen of the Antipodes. Earl did not know what the Antipodes were, but he was beyond doubt the only one of us who could really

ride a horse. I preferred two-legged animals and I myself wasn't entirely sure what the Antipodes were either. McGovern told me to look it up when I got
home, but I soon discovered
the meaning when I found myself in the Outback riding
'cross the desert on a horse with
no legs.

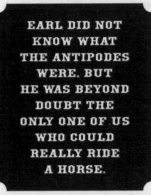

EARL DID NOT
KNOW WHAT
THE ANTIPODES
WERE. BUT
HE WAS BEYOND
DOUBT THE
ONLY ONE OF US
WHO COULD
REALLY RIDE
A HORSE.

The flight to Australia
takes just a trifle longer than
the gestation period of the giant sea turtle, but there are al-
ways at least three full-length movies to divert the
passengers. After each movie, Earl Buckelew, still
wearing his cowboy hat, asked the flight attendants,
whom he consistently called stewardesses, whether
they were male or female, how much time was left in
the flight. Every time he asked it seemed they always
gave him the same answer: "About eight and a half
hours."

A flight of this length gives you a chance to get to
know yourself and your companions a bit more than
you might've wished, but somehow we managed. My
biggest problem was a satanic little baby sitting directly behind us who kept deliberately sneezing upon
me. Tom's road-to-Damascus experience came when
he learned that through some mix-up all that was left

for his lunch was a vegetarian meal. He promptly became highly agitato, but I'm pleased to report that he did settle down after about eight and a half hours. "This is *exactly* what I didn't want to happen," he told the mother of the small infant behind us, whereupon the child maliciously sneezed upon me again.

McGovern made do with many little bottles of vodka, teaching the flight attendants and other privileged passengers how to make a rather arcane drink called a Vodka McGovern. Earl captivated almost everyone he talked to and he never stopped talking once we left the States. They'd never met a real Texan before, one who'd driven an actual horse and wagon, sheared sheep, broken wild horses, built houses with his own hands, and especially one whose own grandfather had been captured by the Comanches. It was a good thing Earl was delighting these people, I thought, because I had pretty well run out of charm, and the baby behind me was now locked in a terrible rictus of unpleasant hiccupping. As for McGovern, he was singing "Waltzing Matilda" with a nun from Syracuse, New York. As they tell you down under: "No worries, mate."

It's probably just as well that the images of our first days in Australia, having percolated within my gray matter department for many years, seem now as only bright pieces of a mosaic in my mind. Even when you're there, there is a faraway, ephemeral, on-the-beach-like quality to Australia that has nothing to do with how heavily monstered you are when you arrive. It's something you see embodied in ancient aboriginal paintings, all of which were created with a series of dots of color. You can see these paintings in some Sydney gallery, but when you fly over the Outback and look down at the landscape, it is precisely and uncannily similar in style and detail to the aboriginal art. Back when the paintings were made, of course, no aboriginal had ever flown in an airplane. In fact, the first person to ever fly an airplane in Australia was an American—Harry Houdini. And about the only feat Houdini never attempted, as far as we know, was to draw aboriginal paintings.

Houdini, however, would definitely not have been challenged by the old hotel we stayed at that first week in Townsville on the Gold Coast. There were no locks on any of the doors. The hotel was a sprawling affair that looked like an old movie set from *Gunsmoke,* with a pub downstairs and a wide veranda encircling the entire structure. Tom and Earl shared the singular honor of having the only room in the place with a bathtub and toilet. I asked Tom whether he would mind monitoring whether the water swirled counterclockwise in the toilet, but, for the moment, he did not seem entirely committed to the project.

Later that night, down in the open-air pub with drunken Aussies throwing darts across the bridges of our noses, McGovern and I were drinking Cascade beer from Tassie (Tasmania) and listening to a British tourist bellyaching about his bad luck. He'd hit a kangaroo with his Land Rover and, not wanting to miss a photo op for the folks back home, had taken the dead body, dressed it up in his jacket, knapsack, and sunglasses, and leaned it against a nearby gum tree. He was taking the kangaroo's photo when

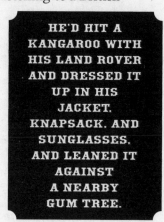

HE'D HIT A KANGAROO WITH HIS LAND ROVER AND DRESSED IT UP IN HIS JACKET, KNAPSACK, AND SUNGLASSES, AND LEANED IT AGAINST A NEARBY GUM TREE.

the animal, apparently only stunned, suddenly bounded away with all his money and his passport.

"Now I'm stuck here broke and getting bitten to death by the bloody flies," he complained.

"Serves you right, mate," said one of the locals. "First you run down one of our 'roos. Next thing, you'll be swattin' our flies."

Before our daunting journey into the vaunted Outback, the Four Horsemen spent a few relaxing days sailing on a forty-two-foot yacht along the Great Barrier Reef. Our hosts, Piers and Suzanne Akerman, were both excellent sailors, which made up for McGovern and myself, who spent most of our time pouring large amounts of Mount Gay rum down our necks and spitting up on the vessel's rather ornate brass compass. At night we watched the Southern Cross, the most beautiful of constellations, roll from one side of the cathedral sky to the other. "Makes you wonder," said McGovern, "what God might've done if He'd had a little money."

"MAKES YOU WONDER," SAID MCGOVERN. "WHAT GOD MIGHT'VE DONE IF HE'D HAD A LITTLE MONEY."

At one point, we dropped anchor and went ashore onto Hamilton Island, considered to be one of the most exclusive travel destinations in the world. Not only did we feed kookaburras by hand, we saw many animals that neither Dr. Seuss nor Earl Buckelew had ever dreamed of. There also appeared, however, to be another kind of animal—a fairly large herd of American businessmen. Tom Friedman confessed to being mildly disappointed. "We've come halfway round the world to this fabled, exotic island," he said, "only to discover an Amway convention."

Several days later, our gutsy little group set out on a secular pilgrimage to the Outback. We had three major objectives in undertaking the trek. One was to prove we could do it. Two was to develop a firsthand understanding of the aboriginal culture. Three was to get the hell away from the Amway convention.

In my twin roles as Virgil leading Dante into the concentric circles of hell, and as official biographer for the Four Horsemen (which was far more tedious), I devised a helpful little list of things a tourist would require to make the trip safely. I also included a list of items an Aborigine would require making the same trip. As an educational service to the reader, the two lists are provided forthwith.

The tourist needs the following things to survive in the Outback: sturdy hiking boots, large canteen with emergency supply of fresh water, first-aid kit, two-way radio, flashlight, tinned foods of every variety under the sun (that's why they need to be tinned), broad-brimmed hat with hole punched in one side of the brim for two-way-radio or walkie-talkie antennae, gun and knife for protection in case you meet someone as crazy as yourself, current map of the area (though nothing's changed since Banjo Paterson wrote the lyrics to "Waltzing Matilda" in 1895), and emergency phone and fax numbers so that you can contact the nearest koala bear with a pager (though, of course, they're not really bears), and antisnake and antispider venom, though if you're bitten by the redback spider it's curtains on opening night. The female redback eats her mate, incidentally, during mating. The male redback, according to Piers Akerman, has a corkscrew-like procreative device. This may account for the female redback's behavior. If you're bitten by the taipan snake, I'm afraid there won't be time to upgrade your software.

The taipan denatures the blood, breaking it down totally and instantaneously. The taipan, the Aussies say, can kill a horse in half a second. That concludes the list of items the tourist needs to survive in the Outback.

The Aborigine's list of necessities is much shorter, of course. In fact, it contains only one item: a stick.

With that same stick, we watched an Aborigine dig up several white larval grubs from under the red dirt of the desert. How he knew where to find them is a mystery locked in past and future aboriginal history, or Dreamtime, as they call it. He popped the thing live into his mouth, placing it headfirst on the back of his tongue so when he swallowed it would crawl downward and not back up.

"We call this witchity grub," he said, offering the sickly white, wriggling object to the Four Horsemen.

We looked at the grub and, I suppose, it looked at us. Tom had been a highly decorated flying hero in World War II. McGovern had been a Marine and had ingested many strange things in his life. When I'd been in the Peace Corps in Borneo I'd once eaten monkey brains. But it was Earl Buckelew, a man who'd never heard of sushi, who was finally brave enough to take the bait, so to speak.

"Care for an after-dinner mint?" I asked him later.

"You know," he said, "it's one of the few things I've eaten in my life that damn sure doesn't taste like chicken."

"YOU KNOW," HE SAID. "IT'S ONE OF THE FEW THINGS I'VE EATEN IN MY LIFE THAT DAMN SURE DOESN'T TASTE LIKE CHICKEN."

In the days ahead, we saw many strange animals and people and did many strange things. We survived climbing Ayers Rock, sacred to the Aborigines (thirty-five climbers at this count have died trying). We saw the famous black swans of Perth. McGovern almost got pecked by a poison parakeet near Darwin. Earl entered and won a sheep-shearing contest in New South Wales. I explored the area where the great Breaker Morant and his horse, Cavalier, had once happily wandered before the Boer War. Tom questioned the authenticity of a cannon that Piers Akerman claimed "fired the first shots in anger in both world wars, Australia being sixteen hours ahead of Texas and probably light-years ahead of Sarajevo where Archduke Ferdinand was assassinated." By the time he'd gotten to how the cannon had started World War II by firing upon a German merchant vessel, the Four Horsemen had cantered off to the nearby town of Robertson, where the movie *Babe* was filmed many years later.

If you go there today and ask for Babe, you'll find that they are tired of fielding questions about their local celebrity. They'll probably tell you, "Sorry, mate. Babe's touring America. He's opening for David Helfgott."

For a country that's roughly the same size as the United States, with only eighteen million people and thirty million kangaroos, both groups of whom happily hop about in the sunshine away from the world's problems, Australia can't be beat. As a vacation paradise, the Four Horsemen give it seven stars, one for every star in the Southern Cross.

But as far as questions about whether the water swirls counterclockwise in toilets or whether dogs circle three times counterclockwise before lying down, I'm afraid you'll have to travel down under yourself to find out. I know, of course, but I can't give you the answers. The Four Horsemen of the Antipodes have taken a sacred, eternal, monastic vow to carry them with us into Dreamtime.

Imus in the Morning

With the decline of the Village People, there aren't many crazy cowboys left in New York City, and Don Imus and I like to keep it that way. We've been friends for almost thirty years, and we spent the first half of that friendship determinedly trying to destroy ourselves. I don't know if you'd call a man who drank a tropical-fish aquarium of vodka every day an alcoholic, but then this is

I DON'T KNOW IF YOU'D CALL A MAN WHO DRANK A TROPICAL-FISH AQUARIUM OF VODKA EVERY DAY AN ALCOHOLIC.

coming from me, the man who snorted the glitter off Loretta Lynn's titter. So who would have thought we'd now be spending our time determinedly trying to help others? Two years ago Don, along with his wife, Deirdre, and his brother, Fred, started a ranch for kids with cancer in that beautiful part of far West Texas some people call New Mexico. Three years ago I helped create the Utopia Rescue Ranch, a never-kill sanctuary for stray and abused animals, located near Kerrville. Don and I do a lot of work and raise a lot of money. It's more satisfying than what we used to do, which was raise a lot of hell.

Don, of course, has millions of radio listeners across the country who tune in faithfully, and you can also see him on MSNBC every weekday morning during the ungodly hours of five to nine. If you don't listen to him, you should. One good reason for doing so is that he's almost constantly promoting me. This does not seem to bother his fans in the slightest. Many of

HIS OUTRAGEOUS BRAND OF HUMOR TENDS TO SAIL DANGEROUSLY CLOSE TO THE TRUTH.

them are outpatients, insomniacs, blue-ball truckers, or spiritual shut-ins, but their numbers are growing, making Don one of the hottest early-morning commodities since my fairy godmother's reheated coffee.

Don's dark, twisted, troublemaking take on life and its myriad shortcomings can be highly addictive, not to mention that his outrageous brand of humor tends to sail dangerously close to the truth. For instance, when President Bush held a press conference a few months back with the pope, Don said, "This ought to be good since neither of them can speak English." He has one of the best horse-manure meters in America for weeding out the phonies, the hypocrites, and the pompous asses among us, as well as an almost encyclopedic knowledge of country music, the blues, history, current events, animals, and of course, people.

Don's version of our long, heroic friendship, however, is somewhat different from mine. He contends that our relationship always seems to warm up considerably whenever I publish a book. This, I can assure you, is not entirely true. Our friendship also heats up whenever I've got a new CD or whenever I'm performing in New York. This should not be terribly surprising. After all, what are friends for?

This past summer I got the chance to visit the Imus Ranch, which is near Ribera, about forty miles east of Santa Fe. It is a four-thousand-acre working cattle ranch for kids terminally ill with cancer and is, among other things, the only vegetarian cattle ranch in the world, proudly sporting a large herd of Longhorns that will live out their days happily, never to be slaughtered. The Imus Ranch is completely organic and serves meals according to a strict vegan diet. I think it's important to be a vegetarian because it's a way of being kind to animals. Of course, it's also a way

of being morally superior to other people. Unfortunately, in the history of the modern world there have been few male vegetarians. The list includes Gandhi, Imus, Dwight Yoakam, Woody Harrelson, Hitler, and myself. Any way you look at it, this is not a particularly strong group.

My only major complaint about the Imus Ranch, in fact, was that I had to sneak out of my building every time I wanted to smoke a cigar. I did, however, get to ride a horse for the first time in many years. Usually, I ride two-legged animals. On this day Don, Fred, Deirdre, ten kids, and I rode for several hours across the rolling hills of New Mexico beneath a majestic mesa. We passed sheep, buffalo, cattle, chickens, goats, donkeys, and an old Western town that would rival any set in Hollywood. Behind the town's facade are the ranch offices, guesthouses, infirmary, and recreation room. Afterward we stopped for lunch at the hacienda, where Don and Deirdre stay with the kids. Don told me the hacienda was his idea, a way of recreating the ambience and spirit of the television series *Bonanza*. I told him that only a mental patient would think of doing something like that. Nevertheless, almost one hundred kids do that each summer, and their numbers are growing. It is important to note that the kids do all of the ranch chores, and they are never treated like cancer victims.

Don may seem like a strange figure to undertake a venture like this. Numerous times I've wondered at his death-defying survival and, indeed, questioned his very sanity. I remember several times when the two of us quite literally saved each other's lives, or maybe we just thought those were our lives. But never have I been prouder of Don or seen him riding taller in the saddle than  that morning last summer when he led those little kids on a trail ride over the hills of New Mexico. All of us, I suppose, could be said to be terminal. It's just that some of us are more terminal than others.

After twenty years of hell-bent self-destruction, Don has risen like a phoenix, and I'm actually beginning to believe that he may have nine lives. His most recent brush with death, however, wasn't all his fault. In June 2000, Don was thrown by a horse—interestingly enough named Destiny—and broke seventeen ribs, punctured a lung, and shattered his clavicle. It took the paramedics two hours to reach him, and they pretty much thought he was a goner. Those two hours,

Don later told his radio audience, were the longest two hours of his life. A listener called in from White Plains, New York, and told Don that the longest two hours of his life was anytime Kinky Friedman appeared on the show.

A Chanukah Story

By sometime around 330 B.C., Alexander the Great had conquered much of the known world, including what we now call the Middle East. Like Napoleon and Hitler, Alexander was short and very much distrusted cats, Jews, and newspapers. Also like Napoleon and Hitler, Alexander did not get to have much fun in life. Soon

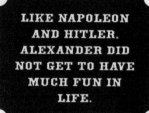

LIKE NAPOLEON AND HITLER, ALEXANDER DID NOT GET TO HAVE MUCH FUN IN LIFE.

after his victories he stepped on a rainbow and spent a great deal of downtime hopping around hell on a pogo stick waiting to be bugled to Jesus, who, of course, would not be born for another three hundred years.

But Alexander left four Greek generals in charge when he blipped off the screen, and one of them, Antiochus, proceeded to get up the colorful, biblical sleeves of the Hebrews rather severely. Antiochus's zealous efforts to make the local populace worship numerous heathen gods became a thorn in the sides of the sons of Abraham and eventually a thorn in the crown of a young Jewish troublemaker named Jesus Christ.

Trouble, indeed, had been brewing for quite some time before Antiochus came up with the brilliant idea of creating the world's first traveling tent revival show, schlepping mobile altars around to the local towns and villages, and making everybody pray to the trendier gods of his choosing. His program met with some success until Mattathias, out on a day pass from the Shalom Retirement Village, saw a Jew sacrificing a pig on one of the heathen altars, then watched him sit there dining placidly on pork tartar. Mattathias, upon observing this rather repellent spectacle, became highly agitato, pulled a large gleaming sword from his flowing mental hospital bathrobe, and skewered the poor fellow, who later turned out to be merely a Reform Jewish businessman from San Diego.

Thus, the true inception of the Chanukah story involves one Jew slaying another, the latter not only being unfamiliar with Jacob's ladder, but definitely never having seen the movie *Babe*.

From there, things got progressively out of touch with the mothership. Mattathias and his five sons, led by Judah the Maccabee a/k/a "The Hammer"—no relation to the former rap star—moved out where the buses don't run. Then, in a series of history's first guerrilla raids, they began to slice up the oppressive Hellenistic Syrian government like a pastrami at Katz's Deli.

THEY BEGAN TO SLICE UP THE OPPRESSIVE HELLENISTIC SYRIAN GOVERNMENT LIKE A PASTRAMI AT KATZ'S DELI.

By 165 B.C., Judah Maccabee and the boys, though greatly outnumbered, had triumphed over Antiochus and were busily at work sweeping the pork rinds, lotto tickets, and Scientology leaflets out of the temple. Antiochus, by this time, was in hell locked in an almost piston-like tandem with Alexander the Great. Every time Alexander began the down cycle on the pogo stick, a precocious Jewish child would hit Antiochus on the head with a small plastic hammer.

The Old Testament tells us that the Lord created the heavens and all the earth in six days. There are those of us, of course, who think He might've taken just a little more time.

That being as it may, the ancient Hebrews, having already gone through a myriad of tedious experiences, were not eager to go against God's sometimes rather overbearing demands. They knew when the fighting is over, whether you've won or lost, it's time for a little prayer huddle. And they remembered the Lord's commandment to Moses upon Mount Sinai: "Take two tablets and go to bed."

And thus it was that quite wisely, instead of coveting their neighbor's ass, or dancing the night away at the Electric Matzoball in Jerusalem, they set about dedicating the temple for prayer.

Hither and thither ran the Israelites in search of oil with which to consecrate the temple by lighting the eternal light. Unfortunately, the Chosen People, following the Lord's wisdom from on high, had chosen to fight and die to live on the one plot of land in the entire Middle East that was totally devoid of oil. And these were the days before OPEC and the big American oil companies had generously made oil available to everyone at reasonable prices. They searched the whole damn temple and all they could find was one little cruse of oil.

That was only enough to last for one night and if you've ever been in an Orthodox synagogue you know that some prayers take longer than that.

Thus it was that a rather Semitic-looking Sancho Panza was dispatched on a donkey in search of oil that wouldn't be readily available there for another two thousand years. History records that the trip took eight days and when Sancho Panza returned with the donkey and the oil, his own ass fell off. But by that time, lo and behold, a miracle had occurred. Instead of lasting for one night, the little cruse of oil had burned for eight days and eight nights. By the pale light of the twentieth century this may not seem like much of a miracle. But the fact is, during this period of biblical history miracles were in very short supply. Indeed, between the time when Moses parted the Red Sea and the time when Jesus turned the water into wine, about the only miracle worth noting was how the people managed without fax machines.

> TODAY IN AMERICA, OF COURSE, CHANUKAH AND CHRISTMAS SEEM TO HAVE MERGED TOGETHER LIKE TWO LARGE CORPORATIONS INTO A COMMERCIALIZED ONE-EYED GIANT TARGETED MAINLY AT CHILDREN.

Today in America, of course, Chanukah and Christmas seem to have merged together like two large corporations into a commercialized one-eyed giant targeted mainly at children, who are the only ones who still may have a ghost of a chance of understanding their true meanings anyway.

Children have a way of grasping things other than presents sometimes, and some of the things they grasp are the things we have forgotten. That is why, given that Santa Claus may have killed Jesus Christ, and given that many American Jews seem ever uncomfortable with who they are, the Chanukah lights somehow continue to shine.

The Germans were able to kill six million but were not able to extinguish the Festival of Lights in Anne Frank's eyes. Deep and dark and bright reflections still dance across the countenance of every child who lights a candle.

In the words of the Hungarian freedom-fighter Hannah Senesch, another child who died defending the dream, "Blessed is the match that kindles the flame."

Robert Lewis Stevenson in Samoa

Robert Louis Stevenson's dark, gypsy eyes always reminded me of Anne Frank's or Elvis's or those of some other hauntingly familiar death-bound passenger of life. They seem to burn with a fever, like embers from that borrowed campfire that provided heat and light to Stevenson's work and to his life. A piece of spiritual trivia, which some may find poignant and some may find stultifyingly dull, is that RLS, during the last five years of his life, possessed the only working fireplace in Samoa. Still, it was not enough to warm his shivering Scottish soul.

What, you might ask, does a cowboy know about Robert Louis Stevenson? What could someone from Texas, where we have wide-open spaces between our ears, possibly hope to accomplish by hectoring the people of Scotland regarding their worst legal scholar and greatest literary lighthouse keeper? We'll see.

WE CAN ALL AGREE THAT HEROES ARE FOR EXPORT.

In the meantime, we can all agree that heroes are for export. In America, for instance, we often have to remind ourselves that JFK is not just an airport, RFK is not just a football stadium, and Martin Luther King is not just a street running through the town. RLS is another set of initials representing a man who aspired to inspire before he expired and, by any account, succeeded beyond his wildest dreams. But just like many New Yorkers hardly notice the Statue of Liberty, Stevenson, perhaps understandably, may be old news to some of you. Yet RLS and the Statue of Liberty have this in common: they've both managed to shine their lights for a long time now, and mankind has managed to follow these beacons through many dark and stormy nights of human history. When we get to the destination, of course, we usually discover that it's only Joan of Arc with her hair on fire.

Writing fiction, I've always believed, is the very best way of sailing dangerously close to the truth without sinking the ship. Stevenson did this as well as anybody when he was alive and, incredibly, in this age of attention deficit disorder, still appears to be going

strong. His spirit seems to transcend the time he never had and
the geography he never got
enough of. I have followed his
footsteps, like a spiritual stalker
in the sands of the South Seas,

I HAVE FOLLOWED HIS FOOTSTEPS, LIKE A SPIRITUAL STALKER IN THE SANDS OF THE SOUTH SEAS.

and everywhere I went it was almost impossible to believe that more than a century has passed since he was
bugled to Jesus.

One of the things that makes Stevenson so endur-
ing and appealing is that before he hit the literary
bigtime, he was, for many colorful, quixotic, heart-
breaking, bohemian years, a jet-set gypsy. Today we
probably would have thought of him as a homeless
person with a sparkle in his eye. Guesthouses he was
summarily thrown out of now bear his name. There is
a rustic area in Northern California where he wandered in deep despair and aching loneliness on a donkey, to paraphrase Leonard Cohen, like some Joseph
looking for a manger. He fell off the donkey into a
canyon bed and, in a weakened state of a fragile life, no
doubt would have perished forgotten if he hadn't been
discovered by two teenage boys. The rustic area is now
known as the Robert Louis Stevenson National Forest.

In 1889, Stevenson, aboard the ninety-four-foot
schooner the *Casco*, departed San Francisco, sailed the
South Seas, and, seven months later, arrived safely at

the harbor in Honolulu. He was beginning the mortal coda of his life. He was thirty-nine years old. He had five years left to live and die in the paradise of his choosing.

THE BEAUTIFUL THING ABOUT INSPIRATION IS THAT IT TRAVELS SO WELL.

The beautiful thing about inspiration is that it travels so well. Stevenson's trip to Samoa had been deeply influenced by his reading of Herman Melville, whom Stevenson, like almost everybody else, assumed was dead. Melville wasn't dead, however; he was just not currently working on a project. He was living out his last days as customs inspector #75 in New York. *Moby-Dick* had already been out for almost forty years and could still only be found in the whaling sections of bookstores. "The important books," Melville had said, "are the books that fail." When he died in 1891, the *New York Times* misspelled his name in its obituary. By then, of course, Stevenson was safely ensconced in Samoa.

In the time he had left, Stevenson's family grew, from his wife, Fanny, her son, Lloyd, and his mother, Margaret, to spiritually encompass almost the entire Samoan people. He had a romantic, some might say misguided, view of the Polynesian race in general. He believed they had the brains, beauty, and spirit of the Ancient Greeks, and that if the world would leave them alone, they would blossom and flourish, becoming the centerpiece on the table of modern civilization. Needless to say, this was not a view shared by the Americans, the British, or the Germans, all of whom had designs on Samoa.

Thus it was that Stevenson found himself supporting a local chieftain named Mataafa, a rebel Robinhood who stood squarely in the way of the powers that be. By this time RLS was already a beloved cultural icon in Samoa. In the mountain of almost a million words he piled up over his lifetime was the Samoan translation of his South Seas story *The Bottle Imp.* It was the first fiction any of the Samoans had ever read in their own language and many of them, perhaps quite correctly, concluded that fiction might just be another way of telling

MANY OF THEM CONCLUDED THAT FICTION MIGHT JUST BE ANOTHER WAY OF TELLING THE TRUTH.

the truth. Many Samoans, indeed, came to believe that the real bottle imp resided in the big safe in the big house on Stevenson's plantation, Vailima.

Stevenson was soon accorded the accolade *Tusitala*, or The Storyteller, a title of great spiritual importance in Samoan culture. There was probably a bit of Lord Jim and a bit of Don Quixote in Stevenson's relationship with the native islanders, not to mention a scrap of Sergeant Pepper's lonely and a shard of Gullible's Travels. Bur this was as it should be, for Stevenson, like the Samoans, was a child-like, romantic, exuberant spirit, and he fit into paradise with the same awkward grace that he fit into the limbo of the white man's world.

RLS might have been the only white man on the planet who believed that Mataafa could be king of the Samoans and that it was important for this to come to pass. It was in this spirit that Stevenson interceded when Mataafa and his followers were captured and imprisoned, gaining the freedom of most of the political prisoners. In gratitude to *Tusitala*, these Samoans, who by nature instinctively despised manual labor, built a road from Apia, the capital, to Vailima. They named it "The Road of the Loving Hearts" and it still stands today as a monument to Stevenson's humanity.

> "THE ROAD OF THE LOVING HEARTS" STILL STANDS TODAY AS A MONUMENT TO STEVENSON'S HUMANITY.

You can drive this road, as I did some years ago, all the way to Vailima, which is now a beautiful museum and library. You can also climb the nearby Mt. Vaea, which I did as well, stand amidst the windy majesty of the glittering Pacific, and commune with the lingering presence of RLS. On the side of his tomb, two verses are inscribed from his poem "Requiem":

Under the wide and starry sky
Dig the grave and let me lie
For glad did I live and glad did I die
And I laid me down with a will.

And these be the words you 'grave for me
Here he lies where he longed to be
Home is the sailor, home from the sea
And the hunter, home from the hill.

There is some special something about the way in which Stevenson passionately interwove his evanescent life with his indelible art that has caused the ensuing embroidery to seem to last forever. Like van Gogh, like Hank Williams, the work defines, sustains, and sometimes destroys its creator. Robert Louis Stevenson's magic is that he gives it to you.

Before he ever got to Samoa, while still in Hawaii, RLS befriended the young Princess Kaiulani and read to her often under their special banyan tree. Kaiulani, another death-bound passenger of life, was the last princess of Hawaii, soon to lose her kingdom, her poetic friend, and her own life at the age of twenty-three as the people of Hawaii and her royal peacocks all cried together. The banyan tree was eventually cut down by the rough hand of progress, but someone was wise enough to save a green branch, which now has grown into a beautiful tree gracing the playground of Princess Kaiulani Elementary School in Honolulu. Beneath the tree is a bronze plaque which bears a verse from a poem Stevenson wrote for her before she left for schooling in Britain.

> Forth from her land to mine she goes,
> The island maid, the island rose,
> Light of heart and bright of face:
> The daughter of a double race.
> Her islands here, in Southern sun,
> Shall mourn their Kaiulani gone,
> And I, in her dear banyan shade,
> Look vainly for my little maid.

HE TAUGHT CROQUET TO THE LEPROSY PATIENTS AT THE GIRLS' SCHOOL.

Stevenson also visited the island of Molokai, shortly after the death of the great holy man Father Damien, whom he very much admired. While there he taught croquet to the leprosy patients at the girls' school. The ephemeral act of teaching croquet to young leprosy patients speaks like a living page torn from Stevenson's own short, afflicted life. As he left for the barge, the young students crowded along the fence to say good-bye. Had he not left then, Stevenson wrote, he would never have been able to.

And in Samoa to this day there is a traditional greeting sometimes given to ship captains and passengers who arrive by sea. Part of the native greeting is a question which translates into English in roughly the following manner: "Is Mr. Robert Louis Stevenson aboard your ship?" There is no easy answer to this metaphysical question. I would like to think, however, that the answer is "Yes, and he always will be."

Catching Heller

The first time I saw Joseph Heller, I fell in love with him. It was the dead of winter in 1981, and though he was almost completely paralyzed in a hospital bed, I could tell by the feverish look in his eyes that he had definitely taken a turn for the nurse. Heller had suddenly been stricken by a little-known but deadly disease of the nervous system, Guillain-Barré syndrome. As his friend

"ANY DISEASE THAT'S NAMED AFTER TWO GUYS HAS GOT TO BE PRETTY DAMNED SERIOUS."

Mario Puzo, author of *The Godfather,* once observed, "Any disease that's named after two guys has got to be pretty damned serious." It was.

Speed Vogel was the friend of Heller's who took care

of him during the time of his illness. Speed had once been roommates with Mel Brooks. Neil Simon, upon observing the ridiculous and hideously ill-suited nature of Speed and Mel's relationship, supposedly based *The Odd Couple* on the two of them. Though Speed was Joe's longtime friend, he came into my life more recently, when I sent out a girl I'd just met to get some Peruvian marching powder and she came back with Speed instead. After trying and failing to chop him up and snort

AFTER TRYING AND FAILING TO CHOP HIM UP AND SNORT HIM, I BECAME FAST FRIENDS WITH SPEED.

him, I became fast friends with Speed. When Joe went into the hospital, Speed took care of Joe's personal affairs and, in effect, became Joe Heller. Speed moved into Joe's house, wore Joe's clothes, used Joe's credit cards, answered Joe's mail and phone calls, and one fateful night invited me down to meet Joe in the hospital.

Even before I met Joe Heller, he had taken a mild interest in me, or at least in my name. Speed Vogel had called and said Heller wanted to use my name and character in a book. The book was to be entitled *Good as Gold*.

"Great," I remember telling Speed. "How much will he pay?"

"That," said Speed, "I can tell you right now. Not a fucking nickel."

"MONEY MAY BUY ME A FINE DOG." I SAID. "BUT ONLY LOVE CAN MAKE IT WAG ITS TAIL."

"Money may buy me a fine dog," I said. "But only love can make it wag its tail. Might as well let him have it. I'd hate to see my character named Kinky Rosenblatt."

Heller, indeed, invoked my name several times in *Good as Gold,* comparing me favorably with Henry Kissinger. Of course, Heller compares everybody favorably with Henry Kissinger. Today, Heller says of Kissinger, "I used to regard him as an evil presence on the planet. Now, I think of him as just another businessman on the make."

I was flying on eleven different kinds of herbs and spices at the time I visited Joe at the hospital, so I brought my friend Don Imus with me, who brought a copy of Heller's second novel, *Something Happened,* for the patient to sign if he was able. All the way to the hospital, Imus kept mumbling over and over again, "This book's a fucking masterpiece." Heller, apparently, agreed with him, because he signed the book and noted that it was his own personal favorite as well.

"It's possible," I said to Heller late last year, "that I'm currently having lunch with the greatest fucking writer in the world."

"It's possible," said Heller, eschewing chopsticks

for a fork. "It would be conceited of me to disagree with you. But I am not sure about that 'fucking' part."

"There must be some British nerd who's a greater author throughout the world than you," I said.

"Nope," said Heller. "Not living. But years ago, just after *Catch-22* came out, I did get to spend some time with Bertrand Russell." As Heller tells it, the great philosopher and pacifist had put the word out that he wanted Heller to visit him at his reclusive lodging in Wales. Joe had long admired Russell, and since the great man was in his nineties, Heller had arranged to see him. "I had just introduced myself to Russell," said Heller, "when he started gesticulating wildly with his cane and shouting 'Go away, damn you! Never come back here again!' "

Heller shook his head in dismay and started to walk to his car when Bertrand Russell's manservant came running out to him. "I'm sorry, sir, but there's been a bit of a misunderstanding. Mr. Russell thought you said 'Edward Teller.' "

For young readers who were jumping rope in the schoolyard at the time and remain ignorant of Mr. Teller's work, he is sometimes remembered as "Eddie" but is more often known as the father of the atomic bomb.

Although Heller claims he looks thirty years younger, he will turn seventy-five on May 1, a certified member of the Shalom Retirement Village People. He's been married to his second wife, Valerie, for more than ten years. We were eating a large steamed flounder in a small restaurant on Mott Street in New York's Chinatown. The fish was almost as big as the restaurant, and the list of questions I planned to ask was almost as big as the fish.

"What about the fucking part of life?" I probed.

"I'm still very sexually attracted and aroused by all

"I'M AN EXPERT FLIRT. I'M WITTY. I'M INTELLIGENT. I'M SENSITIVE. I'M CALCULATING. BUT I NEVER ACT ON IT."

different kinds of women," said Heller. "I'm an expert flirt. I'm witty. I'm intelligent. I'm sensitive. I'm calculating. But I never act on it."

"You hold back for moral reasons?"

"Not really," said Heller. "It's just no longer cost-effective."

No one really knows, of course, how much of a financial pleasure *Catch-22* has been to Heller since it was published in 1961. Publishers, agents, accountants, thirty-year-old investment bankers with three-inch dicks, and Heller himself have all given up trying to keep track of the myriad editions, reprints, versions, and translations of the book, which has probably come closer than any other to being the secular Bible of the world.

"Did you have any inkling as you were writing

Catch-22 that it would someday be a world classic, right up there with Sir Author Conan Doyle, Anne Frank, and *I'm Okay, You're an Asshole?*"

"First of all," said Heller, putting down his fork rather grudgingly, "I operate on two principles: 'Nothing succeeds as planned' and 'Every change is for the worse.' I started the book one morning from left-over dream material. It was 1953, and I was hunched over my desk in an advertising agency. It took eight years to write, and during that time I never called it *Catch-22*. I always wanted to call it *Catch-18*."

"Nothing succeeds as planned," I said.

"I even published the first chapter under the title *Catch-18* in 1955, in a literary quarterly called *New World Writing*. They paid me twenty-five dollars, which wasn't too bad at the time. In the same issue was a chapter from Jack Kerouac's *On the Road,* published under a pseudonym. In 1957, when the book was about half done, I received a contract from Bob Gottlieb [then a rising editor at Simon & Schuster] for fifteen hundred dollars, which also wasn't too bad at the time. There was very little that wasn't too bad at the time.

"Then, in 1961, in order to avoid a conflict with Leon Uris's *Mila 18,* they said the title had to be changed. The publisher felt the public would not buy two books with the number eighteen in the title. I thought about *Catch-11,* but then the movie *Ocean's*

Eleven became popular. My editor and agent and I had been living with *Catch-18* for over five years, and the thought of changing it was a blow to us all.

"One morning that summer, Gottlieb called and said, 'I think I have it. Don't say no until you hear it: *Catch-22.*'

" 'My God!' I said. 'That's it!' "

Catch-22 did not immediately climb into the best-seller heavens like so much effortless German antiair-craft flak. Heller's sardonic, subversive, funny, tragic, timeless tale of a World War II bombardier who's too crazy to want to fly and too sane to want to die isn't now—nor has it ever been—on the *New York Times* best-seller list. Indeed, like Melville, Whitman, and Dickinson, Heller proved to be a somewhat acquired American taste. In the way only a critically crucified author who has risen again can quote a bad review, Heller intoned the following over the flounder from memory with almost boyish glee: "The one I have emblazoned upon my scrotum is from *The New Yorker: 'Catch-22* doesn't even seem to have been written, instead it gives the impression of having been shouted onto paper. Heller wallows in his own laughter and finally drowns in it. What remains is a debris of sour jokes.' "

Heller laughed and began to worry that he might wallow and drown in his wonton soup. For today, thirty-six years and several wars later, Joseph Heller can travel to any literate community in the world and be mobbed. Many of them repeat to him precisely the same phrase: "*Catch-22* is part of my life."

MANY OF THEM REPEAT TO HIM PRECISELY THE SAME PHRASE: "*CATCH-22* IS PART OF MY LIFE."

The dark humor of *Catch-22*—conceived in the shadows of the McCarthy-inspired communist witch-hunt—ensures that the work remains an evergreen in the nearly defoliated forest of war novels written in this century. It is the story of a bomber squadron based on an imaginary island off the coast of Italy during World War II and the struggle of a bombardier, Yossarian, to survive the conflict's last months.

I first read *Catch-22* in 1966, when I worked as a Peace Corps volunteer in the jungles of Borneo. My job was to teach people who had been farming successfully for more than two thousand years how to employ new agricultural methods. Heller's novel proved a spiritual handbook for dealing with local bureaucracy and Peace Corps politics. By the time the bloody mosaic that was Vietnam colored our lives, *Catch-22* almost appeared to have been written about that war. And today, in our sanitized society, Yossarian's one-man crusade against the system resonates more powerfully against the walls and malls of the Nike Village we call the nineties.

"The book would've been considered provocative, even subversive, in the fifties, when I was writing it," said Heller. "By the time it was finally published, most intelligent Americans had come to see the world the way I did and were able to nod in agreement. But governments don't change that much, so the book remains just as pertinent today."

In August 1961, just before the book was published, the nightmarish Cold War reached a flash point with the Berlin crisis. In the minds of many Americans, war with Russia appeared imminent. "My biggest fear in those days was that there would be war with Russia and the book would be banned and the FBI would come and arrest me," Heller said.

Instead, Art Buchwald cabled his congratulations from Paris. Kirk Douglas called from Hollywood. James Jones, Irwin Shaw, and scores of eminent writers heaped praise upon Heller. Tony Curtis telegrammed: "I am Yossarian." Anthony Quinn met Heller at the Sixth Avenue Deli in Manhattan and explained why he should be Yossarian. By the end of 1963, *Catch-22* had become the largest-selling paperback in the country.

"It was extremely exhilarating and exciting," Heller said. "It was in 1962 that I signed the movie deal for $100,000, with payments spread over four years, by which time I felt certain my second novel would be completed. Unfortunately, it took thirteen years."

In 1970, after Mike Nichols's movie version of the book had come out, *Catch-22* sold roughly two million copies in a short time. A year later, Heller received his first big royalty check: $68,000.

"What do you do in your spare time?" I asked.

"I write. If I were seven-foot-two, I'd play basketball. If I were four-foot-two, I'd be a jockey."

"You like racing?"

"No, I'd be a lawn jockey."

"One stupid Barbara Walters–type question: Are you happy?"

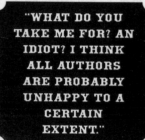

"What do you take me for? An idiot? I think all authors are probably unhappy to a certain extent. They experience inner disruptions or something. Or it could be that everybody's unhappy."

"I know *I'm* unhappy," I said. "The spareribs and black bean sauce are taking forever."

As we walked leisurely up Mott Street toward Little Italy, we spoke of many things: of ships and shoes and Kissingers and Martin Luther Kings. Heller said the only politicians he still trusted were Franklin Delano Roosevelt and former New York mayor Fiorello La Guardia. "The more we know about a candidate, the harder it is to respect him," he said.

As we walked and talked, I wrote many things down in my little private investigator's notebook. I lit up a Cuban cigar and asked him how *Catch-22* played in Cuba.

"*Catch-22* seems to flourish in places where democracy doesn't. It also flourishes in places where capitalism possibly succeeds too well. The entrepreneurial character in *Catch-22,* for instance, Milo Min-

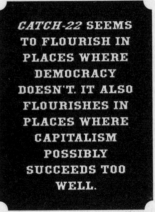

CATCH-22 SEEMS TO FLOURISH IN PLACES WHERE DEMOCRACY DOESN'T. IT ALSO FLOURISHES IN PLACES WHERE CAPITALISM POSSIBLY SUCCEEDS TOO WELL.

derbinder—in the novel he markets chocolate-covered surplus cotton. Today he's probably a Disney executive. Maybe he runs Microsoft. Yossarian lives, but Milo thrives . . . Where were we?"

"Cuba."

"Right. We do very well in places like Cuba. *Catch-22* validates dissatisfaction and solidifies disdain for the people who govern."

"So we could almost say the book is a harbinger of democracy."

"I wouldn't go that far. Why are you belching like an Eskimo?"

"Strictly for health measures. If you keep all bodily gases inside, you'll eventually implode. By the way, is *Catch-22* translated into any Eskimo dialect?"

"No," said Heller seriously. "But you probably couldn't mention another race, language, ethnic group, or geographical locale that hasn't seen a translation over the years."

That evening, Heller and I were sitting at a bar near my hotel. I was drinking Chateau de Catpiss, and Heller had given the following order to the lady bartender, whom he was gazing at but had already deemed not cost-effective. "Extra-dry martini straight up with a twist of lemon in a very, very cold glass, and don't stop at one."

"Obviously, you believe in the sanctity of marriage," I said.

"Well," said Heller, "I married Val in eighty-seven, five years after I left the hospital."

"What did you and Val do during those five years?"

"We lived in sin. Marriage leaves much to be desired, but I've always had a woman to take care of me. My mother, my sister Sylvia; my first wife, Shirley; my second wife, Val; and of course, there were my three years in the army, which turned out to be the biggest mother of them all."

"How does it feel to have universal acclaim?"

"It's better to have it than not."

"But you appear to be a mortal man. Do you believe in God?"

"GOD AND I HAVE A COVENANT. WE LEAVE EACH OTHER ALONE."

"God and I have a covenant," said Heller, working religiously on his second martini. "We leave each other alone."

"In your new book, *Now and Then*, you describe growing up in that colorful, long-ago Jewish and Italian neighborhood of Coney Island, where politicians came from all over for the mandatory photo op, and as you said, 'They smiled because they knew they didn't have to live there.' Has that background influenced your life or your writing?"

"When I was growing up on Coney Island, there were no divorces, no Jewish Republicans, nobody sailed, nobody played golf. Today, I think I'm the only one of that group who's been divorced, but a few of my friends have done well enough to die on golf courses. In *Catch-22* there are forty-four different characters, and none of them are Jewish, except possibly Yossarian. Yes, possibly Yossarian."

I asked Heller to sign a Norwegian translation of *Catch-22* that I had picked up at the airport in Oslo months earlier. He signed the book, the bartender dropped the hatchet, and we walked out onto the twilight sidewalk. Joe told me that as a senior citizen, he could ride the bus forever in New York for seventy-five cents. This was a mode of travel and a visual experience he'd never known, and he was enjoying it like the little boy he'd always wanted to be.

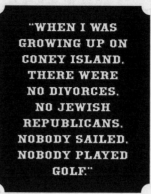

"WHEN I WAS GROWING UP ON CONEY ISLAND, THERE WERE NO DIVORCES, NO JEWISH REPUBLICANS, NOBODY SAILED, NOBODY PLAYED GOLF."

We shook hands on the sidewalk, and after he'd shuffled off toward the bus stop, somewhat cosmically I saw a guy selling Coney Island franks from a pushcart. I bought one for the hell of it and ate the hot dog as I read what Heller had written in the book. "To Richard Kinky 'Big Dick' Friedman: My only authorized biographer. See you tomorrow. Joe Heller. Nov. 20, 1997. NYC."

I don't know if I'll see Joe Heller tomorrow, or if he'll see me or if anybody will see anybody. We all have a covenant with God: that we can't live forever, only fictional characters can do that. Like Cervantes and Don Quixote, Heller has given his crazy bombardier to the world. Yossarian, the holy fool. The thinking man's Jesus Christ. The one we meet first in a hospital bed, just as I'd first met Joe Heller.

Midtown traffic is starting to pick up. Somewhere down the street, an old man is getting on a bus. Or maybe it's just Yossarian disappearing in the mist, dancing toward Sweden.

PART 4

All of the Above

Don't Forget

In the dead of the night I started to write. If the editor wanted more pages, I'd give him more pages. If the agent wanted more action, I'd give her more action. But first, I felt it was necessary to write an homage to Clyde and Fox. As characters, I had them down cold by now, I thought, and certainly I could complete the novel out of my own imagination, which is what every reader would believe it to be anyway. I did not need any longer to faithfully chronicle their ridiculous little hobbies and adventures out of the whole cloth of their existence. They were characters and I was the author. I could make them do or say anything I wanted now. Maybe Clyde had been right all along. Maybe I was destroying them. What an odd occupation I had,

I thought wryly. I was destroying them to create them. But it had to be done. And yet, I missed them. I realized, almost wistfully, that I might never see them again.

I started with Fox, hearing his voice in random past conversations, empathizing with his nuthouse background, getting inside his head. I felt like Faulkner, throwing the story to the winds. I felt like McMurtry, writing two hundred pages of boring shit before I really got going. I felt like J. D. Salinger, who only mixed interpersonally to get inside the heads of real people then cut them out of his solitary life and nailed their hearts and souls to the page with a million typewriter keys. I felt like Fox and I felt crazy like a fox and I felt nothing. So, I said my farewells to Fox by writing a sort of soliloquy in his voice and putting him back in a mental hospital.

> I FELT LIKE MCMURTRY. WRITING TWO HUNDRED PAGES OF BORING SHIT BEFORE I REALLY GOT GOING.

"A mental hospital is not always as romantic a place as it's cracked up to be. You always think of Ezra Pound or Vincent van Gogh or Zelda Fitzgerald or Emily Dickinson or Sylvia Plath or someone like that. Not that all the above-mentioned people resided in mental hospitals. All of them probably belonged there, but so do most people who don't reside in mental hospitals. I know Emily Dickinson never went to a mental hospital, but that's just because she never went anywhere except, of course, for brief walks in her garden with her dog, Austin. If she'd ever gone into a mental hospital and talked to the shrinks for a while they never would have let her out. She might've done some good work there but that would've been her zip code for the rest of her life. Now you take van Gogh, for example. He lived in one with a cat and did some good work there. They put him in for wearing lighted candles on his hat while painting *Night Cafe*. Today, the arbiters of true greatness, Japanese insurance companies, have determined that his work is worth millions. Sylvia Plath I don't know too

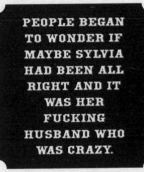

PEOPLE BEGAN TO WONDER IF MAYBE SYLVIA HAD BEEN ALL RIGHT AND IT WAS HER FUCKING HUSBAND WHO WAS CRAZY.

much about except she wrote good prose and maybe some great poetry and then she put her head in an oven and killed herself but by then it was too late for her to reside in a mental hospital. Everybody thought she was crazy for many years until her husband's second wife also croaked herself and then people began to wonder if maybe Sylvia had been all right and it was her fucking husband who was crazy. I mean to have two wives conk on you like that, I mean each one topping herself on your watch, pretty well indicated

'Scuse Me While I Whip This Out   **157**

to most people outside of mental hospitals that if that husband wasn't crazy there was something wrong with him. Now Ezra Pound I don't know a hell of a lot about except he hated Jews and still managed to do some pretty good work in wig city. Hitler and Gandhi, both of whom probably belonged in wig city, for different reasons, no doubt, somehow managed to avoid the nuthouse circuit. They each did, however, spend a bit of time in prison, which in some ways is not as bad as being in a mental hospital except that you come out with an asshole the size of a walnut. In a sense Hitler and Gandhi, who may well represent polar opposites of the human spirit, each found himself in prison where the absence of freedom and the distance from their dreams may have contributed to their achieving some pretty good work. Hitler, who hated Jews almost as much as Ezra Pound, wrote *Mein Kampf,* which was almost immediately translated into about fourteen languages and would have made him quite a favorite at literary cocktail parties if he'd been willing to stop there. Unfortunately, he couldn't hold a candle to Anne Frank. Gandhi, who spent his time in prison listening to a South African mob singing 'We're gonna hang ol' Gandhi from a green apple tree,' did some scribbling of his own but mostly realized that he was tired of London yuppie lawyer drag and it was time for visions and revisions both sartorially as well as spiritually. But God only knows how Hitler and Gandhi, who were both interesting customers, would have fared had they been incarcerated in mental hospitals instead of prison. As it was, each man found himself creating and writing in the calaboose, something that almost never

BUT GOD ONLY KNOWS HOW HITLER AND GANDHI, WHO WERE BOTH INTERESTING CUSTOMERS, WOULD HAVE FARED HAD THEY BEEN INCARCERATED IN MENTAL HOSPITALS INSTEAD OF PRISON.

happens in a mental hospital because shrinks are constantly prescribing meds which keep you invariably, perpetually, hopelessly lost. Speaking of lost, Zelda Fitzgerald certainly qualifies in that category and technically, I suppose, she was confined to a 'sanatorium' which was not truly a mental hospital if you want to be a purist about it but no doubt still probably had a sign in the lobby that read: THIS IS TUESDAY. THE NEXT MEAL IS LUNCH. She'd been drinking a lot of her meals evidently and so they'd put her in this sanatorium in Asheville, North Carolina, or maybe it was Asheville, South Carolina. I always get those two states mixed up. Where are the Wright Brothers when you need them? Anyway, the irony of the whole situation was that the sanatorium was in Asheville and the place burned down one night with Zelda and a fairly good-size number of other no-hopers inside. I've wondered why God so often seems to send fires and other catastrophes to sanatoriums and mental hospitals. It's kind of like swerving to hit a school bus. But all that being as it may, it's just ironic I thought that the sanatorium burned down and that it was in Asheville. But before Zelda came along to screw things up I was commenting on the fact that mental hospitals are far more sad and sordid places than you'd think, seeing as all these colorful, fragile, famous, ascetic people populate them. I mean it isn't all van Gogh and his cat. I mean there are men following you with their penises shouting, 'Am I being rude, mother?' in frightening falsetto voices. People in mental hospitals are shrieking like mynah birds all the time. Or masturbating. Now Dylan Thomas was a good one. He used to masturbate a lot but I don't think they ever put him in a mental hospital though God only knows he belonged there. And speaking of God only knows, Brian Wilson undoubtedly belongs there too, except what would happen to the Beach Boys if you put Brian Wilson in the nuthouse? I mean the only one of those guys who was really a surfer was Dennis Wilson. And

Kinky Friedman

you know what happened to him? He drowned. Ah well, the channel swimmer always drowns in the bathtub, they say. But I suppose I've come pretty far afield in this tawdry little tale which the shrinks

THERE ARE MEN FOLLOWING YOU WITH THEIR PENISES SHOUTING. "AM I BEING RUDE, MOTHER?"

would assuredly call a rambling discourse. But if getting to the point is the determinant of whether or not you're crazy, then half the world's crazy. Trouble is it's the wrong half. I mean whoever said anything important by merely getting to the point? Did guys like Yeats and Shelley and Keats, who, by the way, all belonged in mental hospitals, ever get to the point? I mean what's the point of getting to the point? To show some shrink with a three-inch dick that you're stable, coherent, and well-grounded? And I haven't even gotten to Jesus yet. Sooner or later everybody in a mental hospital gets around to Jesus and it's a good thing that they do because I'll let you in on a little secret: Jesus doesn't talk to football coaches. He doesn't talk to televangelists or Bible Belt politicians or good little churchworkers or Christian athletes or anybody else in this god-fearing, godforsaken world. The only people Jesus ever really talks to are people in mental hospitals. They try to tell us but we never believe them. Why don't we, for Christ's sake? What have we

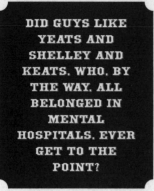

DID GUYS LIKE
YEATS AND
SHELLEY AND
KEATS. WHO. BY
THE WAY. ALL
BELONGED IN
MENTAL
HOSPITALS. EVER
GET TO THE
POINT?

got to lose? Millions of people in mental hospitals who say they've talked to Jesus can't all be wrong. It's the poor devils outside of mental hospitals who are usually wrong or at least full of shit and that's probably why Jesus never talks to them. Anyway you can probably tell by the fact that I'm not employing any paragraphs and the fact that this little rambling discourse tends to run on interminably that this looks like a mental hospital letter itself. If that's what you think, you're right, because I am in a fucking mental hospital as I'm writing this tissue of horseshit and it's not one of those with green sloping lawns in that area between Germany and France that I always forget the name of. Hey, wait a minute! It's coming to me. Come baby come baby come baby come. Alsace-Lorraine! That's where the really soulful mental hospitals are. Unfortunately, I'm writing this from a mental hospital on the Mexican-Israeli border and I'm waiting for a major war to break out and they don't have any green sloping lawns. They don't even have any slopes. All they have is a lot of people who talk to Jesus, masturbate, and don't believe they belong in here. It's not a bad life, actually, once you get the hang of it, unless of course you hang yourself, which happens here occasionally usually on a slow masturbation day. Anyway, the reason I'm telling you all this is that I really don't belong here. I've told the doctors. I've told the shrinks. I've even told a guy who thinks he's Napoleon. The guy's six feet tall, weighs two hundred and fifty pounds, and he's black, and he thinks he's Napoleon. I probably shouldn't have told

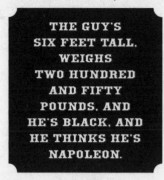

THE GUY'S SIX FEET TALL, WEIGHS TWO HUNDRED AND FIFTY POUNDS, AND HE'S BLACK, AND HE THINKS HE'S NAPOLEON.

him in the first place. The other day a woman reporter came in here from the local newspaper to do some kind of exposé on the place and she interviewed some of the patients and one of them was me. I told her I was perfectly sane and I didn't belong in here. She asked me some questions and we chatted for a while and then she said that I sounded really lucid and normal to her and she agreed that I really didn't belong in here. Then she asked me since I seemed so normal what I was doing here in the first place and I told her I didn't know I just woke up one day and here I was and now the doctors won't let me out. She said for me not to worry. She said when she finished her exposé on my condition, these doctors would have to let me out.

Then she shook my hand and headed for the door. About the time she put her hand on the doorknob, I took a Coke bottle and threw it real hard and hit her on the back of the head.

" 'Don't forget!' I shouted."

Wild Man from Borneo

Many years ago, in a faraway kingdom called The Sixties, when doctors drove Buicks and ecstasy couldn't be bought, there lived a man named John F. Kennedy. One day he stood on the lawn of the White House, pointed at a group of ragtag young Peace Corps volunteers, and said, "You are important people." And, indeed, time has proven the wisdom of his words. Forty-one years and more than one hundred countries later, the Peace Corps is a shining example of Americans working for the good of the world.

Little did I realize in 1965, as I drank coffee at the Night Hawk restaurant on the Drag in Austin and contemplated joining the late J.F.K.'s dream team, that I would soon be eating monkey brains in the jungles of Borneo. At the time, I was a Plan II major at the University of Texas. Plan II was a highly advanced liberal arts program mainly distinguished by the fact that every student had some form of facial tic. There was nothing practical about graduating with a degree in Plan II. About all you could do with it was leave town with the carnival or join the Peace Corps. After much soul-searching, I opted for the one that would look best on my résumé.

I soon found myself in Syracuse, New York, in about twelve feet of snow, in Peace Corps training. My only friend was a guy named Willard who smoked nonfiltered Camels and, during the first night's mixer, promptly ran out onto the dance floor and bit a woman on the left buttock. Since these were the good old days before political correctness, Willard was not sent home ("deselected" was the term then in use) and went on to distinguish himself setting up a law school in Africa.

> DURING THE FIRST NIGHT'S MIXER, HE PROMPTLY RAN OUT ONTO THE DANCE FLOOR AND BIT A WOMAN ON THE LEFT BUTTOCK.

I did not fare quite as well as Willard, however. As part of my training, the Peace Corps sent me on a two-week "cultural empathy" junket to Shady Rill, Vermont, where I lived with a family so poor that they brushed their teeth with steel wool. After returning to Syracuse, I learned Swahili and was interrogated at great length by Gary Gappert, a supercilious, pipe-smoking psychologist who

I LIVED WITH A FAMILY SO POOR THAT THEY BRUSHED THEIR TEETH WITH STEEL WOOL.

felt that I might not be fully committed to the goals of the Peace Corps because I had a band back in Texas called King Arthur and the Carrots. Soon, much to my chagrin, I was the one the Peace Corps had chosen to be deselected.

I traveled about the country like a rambling hunchback, hitchhiking from place to place, singing Bob Dylan songs at truck stops. The truckers were not pleased. They enjoyed my behavior only marginally more than Gary Gappert had. Yet I had not abandoned my dream, and eventually I landed at another Peace Corps training program, this time in Hilo, Hawaii, where I was, at long last, hailed as a golden boy. It was also where I learned Malay, a language I can now speak only when I'm walking on my knuckles.

Ultimately I was sent to Borneo, where I wore a sarong, built compost heaps, and earned eleven cents an hour as an agricultural extension worker. My job was to teach people how to keep their heaps from falling over on top of the Kinkster. Somehow I managed to avoid the fate of one of my co-workers, who

'Scuse Me While I Whip This Out

had to be airlifted out of his hut by a shrink in a helicopter.

By the time Martin Luther King and Bobby Kennedy had been assassinated, I'd gone native. I'd taken to spending a lot of time at a Kayan longhouse fairly deep in the *ulu,* or jungle, up the Baram River from the little town of Long Lama. The Kayans were a spiritual people, but they were also rather serious party animals. They had a traditional combo that might have even been stronger than a John Belushi cocktail. It called for chewing betel nut until your lips turned bloodred, smoking an unidentifiable herbal product in a jungle cigar, and then drinking a highly potent homemade rice wine called *tuak* that would have made George Jones jealous. The Kayans, like a tribe of persistent mother hens, would push this combination on every guest, and it was considered extremely bad form to turn down their offering. Accepting their largesse, however, would invariably lead to projectile vomiting. The Kayans had no perceptible plumbing, of course, so you'd simply vomit through the bamboo slits of the porch, or *ruai.* If, after being sick, you continued drinking *tuak* with them, the Kayans considered you a man and, even

> THE KAYANS HAD A TRADITIONAL COMBO THAT MIGHT HAVE EVEN BEEN STRONGER THAN A JOHN BELUSHI COCKTAIL.

more important, a friend. The only time the Kayans found my behavior socially unacceptable was once when, after an extended harvest celebration, I accidentally vomited on the chief.

As a Peace Corps volunteer, my mission was to preserve the culture as much as possible while attempting to distribute seeds downriver. In two and a half years the Peace Corps failed to send me any seeds, so I was eventually reduced to distributing my own seed downriver, which led to some rather unpleasant repercussions. I was well aware that the Kayans, though now a gentle people, had once been headhunters, and I did not

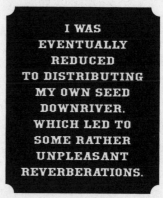

I WAS EVENTUALLY REDUCED TO DISTRIBUTING MY OWN SEED DOWNRIVER. WHICH LED TO SOME RATHER UNPLEASANT REVERBERATIONS.

want an atavistic moment to occur in which my skull might take its place along with dozens of others in the hanging baskets that festooned the *ruai*. But while I supported the indigenous culture, the missionaries were constantly at work to destroy it. They encouraged the Kayans to cut off their long hair, throw away their hand-carved beads, and dance around the fire singing "Oh! Suzanna." I've got nothing against "Oh! Suzanna"—only against the missionaries who told the people to bow their heads and pray long enough so that when they looked up, their traditions were gone.

In a few short years, I was gone too. But all Peace Corps volunteers keep a little town or a little tribe deep in their heart, though they may have left it many years ago and many miles away. I remember fishing at night by torchlight with the Kayans in the Baram River in a small wooden boat called a *prahu*. Everybody got drunk on *tuak* and had a great time, though the Kayans never caught any fish. Of course, that wasn't their intention.

I also remember the coffee-colored river. It seemed to flow out of a childhood storybook, peaceful and familiar, continue its sluggish way beneath the moon and the stars and the tropical sun, and then pick up force and become that opaque uncontrollable thing roaring in your ears, blinding your eyes, rushing relentlessly round the bends of understanding, beyond the banks of imagination.

Blowin' Smoke

When my editor suggested that I write an essay on the subject of cigars, I thought it was a remarkably ill-advised idea. Few people in this world like cigars, and fewer still like the people who smoke them. Most men wouldn't touch a cigar with a barge pole (notwithstanding the fact that some cigars today are almost the size of a barge pole). The vast majority of women despise cigars more than they hate cockroaches. And children, of course, are the worst of all, waving their precious little hands in front of their precious little noses in supermarkets at the mere sight of even an

CIGAR SMOKERS ARE ONE OF THE MOST REVILED AND PUT-UPON MINORITIES IN THE COUNTRY.

unlit cigar. This never fails to irritate the Kinkster. Sometimes I'd like to poke them in the ribs with a barge pole.

Yet that is exactly why I decided to write this piece. I see it as an educational service for the pompous, politically correct, humorless, constipated spiritual bullies in our pathologically health-conscious, civic-minded society who tend to treat cigar smokers like biblical lepers. In fact, cigar smokers are one of the most reviled and put-upon minorities in the country. Fortunately, most of us are large, abrasive, loud-mouthed, wealthy men, and we can handle any abuse that may come our way from precious little kids in supermarkets.

Though the opinion is not widely shared by the scientific community, I have always believed that cigars are good for you. During a crisis, nonsmokers often stand passively by, inhaling passive smoke into their passive lives. Under the same circumstances, cigarette smokers usually break into a panic and jump from the windows of their basement apartments. However, when cigar smokers encounter a trying moment, they often appear to be calmly directing some cosmic symphony that only they can hear. Their attitude can best be described as "Let the lava flow." This mind-set, I contend, lowers the blood pressure and reduces stress, not to mention that it's always fun to enjoy yourself while irritating others. This last part, I maintain, is one of the true secrets to the cigar smoker's vaunted longevity.

Thomas Edison, for instance, who reportedly smoked eighteen cigars a day, lived to be eighty-four, mainly by blowing smoke at all the people who ridiculed his inventions. Mark Twain, who smoked up to forty cigars a day, said, "If smoking is not allowed in heaven, I shall not go." He was seventy-four when he took up celestial cigars and began puffing halos instead of smoke rings. And then there was Winston Churchill, who almost always had a cigar in his mouth during the war years. He even requested a special oxygen mask so that he could smoke during high-altitude flights. He lived to be ninety.

Sigmund Freud, the first man to realize that smoking cigars was a highly suggestive oral fixation, nonetheless was rarely without a cigar himself and lived to be eighty-three. In a rather unpleasant turn of events, I myself was unceremoniously tossed out of the Freud museum in Vienna, Austria, three years ago. The woman at the desk, who looked and behaved like a ferret with earrings, did not take kindly to my smoking a cigar in the museum, which was dedicated to a man whose very life revolved around cigars and the sick things they might represent in the subconscious mind. I was pointing this out when she summoned a phalanx of former U-boat commanders who escorted me to the door as I chanted repeatedly in rhyming verse, "Sigmund Freud would be annoyed!"

As for myself, I smoke as many as twelve cigars a day, and I expect to live forever. Of course, I don't inhale. I just blow smoke at small children, flowering plants, and anybody who happens to be jogging by. At fourteen I experimented with Swisher Sweets and Rum Crooks before moving up to the large black phallic symbols I smoke today.

> A MAN WHOSE VERY LIFE REVOLVED AROUND CIGARS AND THE SICK THINGS THEY MIGHT REPRESENT IN THE SUBCONSCIOUS MIND.

I believe that the plight of the cigar smoker is symptomatic of something seriously askew in the priorities of this great country. Let us say, for example, that three men walk into a restaurant in California. The first is wearing a cowboy hat and smoking a cigar the size of a large kosher salami, the second is carrying an Uzi, and the third, having just returned from a yoga seminar in Utah, is naked and attempting to moon as many patrons as possible. It should come as no surprise that the maître d' and the customers would almost instantly confront the cowboy and have him arrested.

The other two men would be left wondering what the hell happened. Maybe the police would come back later and ask to see their screenplays.

I can only hope for the day when the role of the cigar in our society will no longer be relegated to the ash heap of history. There is something about a cigar that provides a person with a certain spiritual ballast, a measure of peace of mind, and a handy device for keeping the world at bay. Cigar smoking is more than a habit, more than a Freudian slip of the tongue around merely another tobacco product. Smoking cigars is a secular religion, a way of life for those men (and women) who are not afraid to live.

SMOKING CIGARS IS A SECULAR RELIGION. A WAY OF LIFE FOR THOSE MEN (AND WOMEN) WHO ARE NOT AFRAID TO LIVE.

One such person was Ernie Kovacs, a major cigar smoker and a much-loved pioneer of iconoclastic comedy on late-night television. David Letterman and many other comedians continue to borrow heavily from Kovacs's material, philosophy, and style. We'll never know how much he might have contributed to a culture in need of a laugh because his life was cut short in 1962, at the age of forty-two, when he crashed his car. According to legend, the police found Kovacs slumped behind the wheel, his ever-present cigar in one hand and an unlit match in the other. There was widespread speculation that his trying to light the cigar might have caused the accident. What I think happened, however, is that after the crash, knowing he had only moments to live, his last act on earth was to attempt to light his cigar.

Cliff Hanger

On the night of December 17, 1998, I clung precariously to life, sanity, and a sheer cliff side overlooking an angry sea. My only companions were lizards, iguanas, and the pale light of the Mexican moon shining like a white, luminous buttock in the mariachi sky. I'd been staying just outside Cabo San Lucas at the mansion of my friend John McCall, had taken a solitary predinner power walk on the beach, and had been swept out into the ocean by a freak wave. The undertow, which killed a person that same night, swept me hundreds of yards away from the beach and deposited me at the base of a steep cliff. I tried to scramble up, but I found myself trapped between the tide and the darkness. As the water pounded ever

higher along the black, crumbling landscape, intimations of mortality flooded my fevered brain. Like Arafat after his plane crash in the desert, I vowed to be a different kind of person if I survived. I thought of my mother and my cat, both of whom had gone to Jesus. I realized that I might now be seeing them sooner rather than later.

I also thought of what a bothersome housepest I'd turned out to be for my generous host, John McCall. McCall, who is also known as the Shampoo King from Dripping Springs, could afford to be generous. He runs the beauty supply company Armstrong McCall and, as he once told me, is a "centimillionaire." For those of us who can't count that high, it means McCall

is worth a hundred million dollars. Even with inflation, that's not too bad. "Shampoo," says McCall, "makes people feel good about themselves."

As I held on desperately to the cliff, I took some comfort in knowing that McCall had more money than God. There was no way, I figured, he would allow his favorite Jewboy to die an untimely death without launching a land, air, and sea search. As I shivered in the darkness, I listened for helicopters that never came and resolved that if McCall wasn't thinking of me, I would think of him, thereby goosing him into action.

I thought of how McCall had been through hell a couple of times and come out laughing at the devil. In 1990 he himself had almost gone belly-up. Medical experts diagnosed him with deadly lymphoma and pointed the bone at him, giving him only weeks to live. Yet incredibly, McCall had a dream aboard an airplane in which the cancer turned to water and disappeared. When he went in for his next examination, the cancer was, in fact, gone. The doctors had never seen anything like it, but of course, that's what they usually say. Either that or you'll never walk again. McCall did, indeed, beat the first cancer, and when it returned years later, he beat it again. In the interval, just to keep in practice, he survived a plane crash in Alaska.

> **MCCALL HAD A DREAM ABOARD AN AIRPLANE IN WHICH THE CANCER TURNED TO WATER AND DISAPPEARED.**

Now, as I clung to the cliff, soaking wet and shivering in the predawn moonscape, I hoped some of McCall's vaunted luck would rub off on me. What I didn't know that fateful night was that McCall was not really looking for me at all. It wasn't until later that morning, when he discovered my passport, cash, and cigars still in my luggage, that he swung into ac-

'Scuse Me While I Whip This Out

tion. By this time I was dehydrated, delirious, and waving frantically to every fishing vessel I could see, many of whom waved back cheerfully or held up their catch of the day. Because I was trapped, ironically, on a private beach beneath luxury homes, they had no idea that the date on my carton was rapidly expiring. But McCall knew how to launch a major campaign. Soon the FBI, CIA, and DEA were involved, Don Imus's private jet was standing ready in New York, P.I. Steve Rambam had been consulted, and a large blowup of my passport photo, which strongly resembled a Latin American drug kingpin, could be seen on flyers on every telephone pole, hotel, hospital, morgue, and whorehouse in the greater Cabo area.

I, of course, knew none of this. I just kept concentrating on McCall, hoping I was getting through. I visualized a world traveler with a large wad of cash he calls "whip-out." I pictured a mysterious magnate who happily worked as a roadie selling T-shirts on my recent concert tour of Europe. A man who makes huge donations to worthy causes almost always under the name Anonymous. A man who invites the Dallas

Cowboys cheerleaders to his birthday parties, which he often doesn't attend himself. A man with a gazillion-dollar home outside Austin that is known as the Taj McCall. Yet money, I reflected, never seems to make people happy. As McCall himself once told me, "Happiness is a moving target."

Late in the afternoon, my hopes were fading. If I survived, I vowed, they could give me a goat's head and I'd dance all night. Once again I began stumbling upward, lost  in the rocky landscape, trying to find a way to the top of my upscale death trap. Suddenly, while climbing a steep ledge, I was miraculously plucked from my precipice by an intrepid band of Mexicans who were rappeling downward. They had been working on Sly Stallone's house, and McCall had commandeered them. Fortunately, they knew exactly where to look: The same thing had happened to another person just weeks earlier. Sly was not home at the time, but McCall was waiting at the top with a warm hug and cold *cerveza*. To paraphrase my father, it felt almost good to be alive.

That night, after *ocho* tequilas, I asked McCall what took him so long. He explained that he didn't take my disappearance seriously at first. McCall remembered a conversation the two of us had had sev-

'Scuse Me While I Whip This Out   **181**

eral years earlier when we toured the Australian Outback. We had discussed how easy it would be for a person to disappear if he wanted to. McCall, in other words, was convinced that my absence was staged, quite possibly as some kind of publicity stunt. I'd never been averse to a little publicity, of course. I just didn't want to die from exposure.

Some days later, without pulling any punches, McCall finally revealed to me the thing that might have been the toughest blow of all. "The real tragedy," he said, "is that you were fifteen minutes away from making CNN."

Talent

ike the tides, the seasons, and the Bandera branch of the Jehovah's Witnesses, the Texas Book Festival is coming around again, allowing us to meet authors we love, hate, or very possibly, find a little ho-hum. I always look forward to the book festival because it provides me with the spiritual soapbox to give advice to other authors, an audience that, predictably, has never learned to listen. Conversely, I've never learned to pull my lips together, so the system works. My advice to authors, and the misguided multitudes who want to be authors, is a variation on a truthful if sometimes tedious theme. "Talent," I tell them in stentorian tones, "is its own reward. If you're unlucky enough to have it, don't expect anything else." These wise words, of course, come from a man who's spent his entire professional career trying to eclipse Leon Redbone.

My theory is that in all areas of creative human endeavor, the presence of true talent is almost always the kiss of death. It's no accident that three of the people who were tragically forced into bankruptcy at the end of their lives were Edgar Allan Poe, Oscar Wilde, and Mark Twain. It's no fluke of fate that Schubert died shortly after giving the world the *Unfinished Symphony.* You probably wouldn't have finished it either if you had syphilis and twelve cents in your pocket. Or how would you like to have died at age twenty-nine in the backseat of a Cadillac? If you're Hank Williams, that's what talent got you. But what *is* talent? And why would anyone in his right mind want it? As Albert Einstein often said, "I don't know."

YOU PROBABLY WOULDN'T HAVE FINISHED IT EITHER IF YOU HAD SYPHILIS AND TWELVE CENTS IN YOUR POCKET.

In fact, talent is such a difficult quality to identify or define that we frequently end up losing it in the lights, relegating it at last to the trash bin, the cheap motel, the highway, the gutter, or the cross. Indeed, if you look with an objective eye at the *New York Times* bestseller list, the *Billboard* music charts, and the highest-rated network TV offerings, the one thing they seem to have in common is an absence of original creative expression, i.e., talent.

My editor says I'm one of the most talented writers he knows. The problem is that even if I have talent, I don't know what it is—and if I did, I'd get rid of it immediately. Then I'd be on my way to vast commercial success. Talent, however, is a bit like God; you never see it, but there are moments when you're pretty sure it's there. So because I can't clinically isolate it, I'm stuck with all my wonderful talent, and the most practical thing I can do is start looking for a sturdy bridge to sleep under or a gutter in a good neighborhood.

If you have a little talent, you're probably all right. Let's say you're good at building birdhouses or you play the bagpipes or, like my fairy godmother, Edythe Kruger, you do an almost uncanny impersonation of the duck on the AFLAC commercial. These kinds of narrow little talents have never harmed a soul, or kept anyone from living a successful, happy life. It's when you're afflicted with that raw shimmering, innate talent— talent with a big "T"—that you can really get into trouble. Remember that Judy Garland died broke on the toilet. Lenny Bruce also died broke on the toilet. Jim Morrison, just to be perverse, died fairly well financially fixed at the age of twenty-seven in a Paris bathtub. Elvis also died on the toilet, but definitely he wasn't broke. Along with a vast fortune, he had well over a million dollars in a checking account that drew no interest. Who cares about money, he figured, when you've got talent? I myself was a chess prodigy, playing a match with world grandmaster Samuel Reschevsky when I was only seven years old. It's been downhill from there. These days I find myself constipated most of the time and I never take a bath.

> **THEY SAY IT TAKES MORE TALENT TO SPOT TALENT THAN IT DOES TO HAVE TALENT.**

They say it takes more talent to spot talent than it does to have talent. Conversely, it's easy to know when it isn't there, although someone without talent rarely notices its absence. Some friends of mine had a band once, and they went to audition for a talent scout in his office. The talent scout said, "Okay, let's see what you can do." The leader of the band began to pick his nose while playing the French horn. Another guy started beating out the rhythm on his own buttocks while projectile vomiting on the man's desk. The other two members of the band jumped simultaneously onto the desk and began unabashedly engaging in an act too graphic to describe here. "I've seen enough," shouted the talent scout in disgust. "What do you call this act anyway?" The French horn player stopped playing the instrument and stopped picking his nose. "We call our-selves," he said, "the Aristocrats."

Another example of what might help define talent takes us back to Polyclitus, the famous sculptor in ancient Greece. Polyclitus, it is said, once sculped two statues at the same time: one in his living room, in public view, and one in his bedroom, which he worked on privately and kept wrapped in a tarpaulin. When visitors came by, they would comment on the public work, saying, "The eyes aren't quite right," or "That thigh is too long," and Polycli-tus would incorporate their suggestions into his work. All the while, however, he kept the other statue a secret. Both works were completed at about the

SO WHAT CAN YOU DO IF YOU DON'T HAVE TALENT? RELAX AND ENJOY IT.

same time and were mounted in the city square in Athens. The statue that had been designed by com-mittee was openly mocked and ridiculed. The statue he'd done by himself was immediately proclaimed a great transcendental work of art. People asked Poly-clitus, "How can one statue be so good and the other so bad?" And Polyclitus answered, "Because *I* did this one and *you* did that one."

So what can you do if you don't have talent? To paraphrase Claytie Williams, you can relax and enjoy it. Any no-talent fat boy can make it to the top of the charts, but it takes real talent, like that of the brilliant American composer Stephen Foster, to die penniless in a gutter on the Bowery. But with or without talent, you might ask, how can hard work and perseverance pay off in the creative field? Why are you asking me? Who the hell knows? In this day and age, just as the tortoise is finally crossing the finish line to win the race, he'll very likely see three men in suits and ties, standing there with their briefcases. "Hello," they'll say. "We're the attorneys for the hare."

The Navigator

Because I'm the oldest living Jew in Texas who doesn't own real estate, and given my status in general as a colorful character, there are those who profess to be surprised that I ever, indeed, had a father or a mother. I assure you, I had both.

For many years my parents owned and directed Echo Hill Ranch, a summer camp near Kerrville where I grew up, or maybe just got older. I remember my dad, Tom Friedman, talking to all of the campers on Father's Day in the dining hall after lunch. Each summer he'd say essentially the same words: "For those of you who are lucky enough to have a father, now is the time to remember him and let him know that you love him. Write a letter home today." Many years have passed since I last heard Tom's message to the campers, but love, I suppose, has no "sell by" date.

When my father was a young boy growing up in the Chicago of the late twenties, his first job was working for a Polish peddler. The man had a horse and cart that was loaded up with fruits and vegetables, and Tom sat on the very top. Through the streets and alleys of the old West Side they'd go, with the peddler crying his wares in at least five languages and my father running the purchases up to the housewives who lived on the top floors of the tenement buildings. There were trolley cars then and colorful clotheslines strung across the sooty alleys like medieval banners. My father still remembers the word the peddler seemed to cry out more than any other. The word was *kartofel*. It is Polish for "potato."

In November 1944 my mother, Minnie, gave birth to me in a manger somewhere on the south side of Chicago. (I lived there one year, couldn't find work, and moved to Texas, where I haven't worked since.) And all this time my father was far away fighting for his country and his wife and a baby boy he might never see. Tom was a navigator in World War II, flying a heavy bomber for the Eighth Air Force, the old B-24, also known as the *Liberator,* which, in time, it was. Tom's plane was called the *I've Had It.* He flew thirty-five successful missions over Germany, the last occurring on November 9, 1944, two days after he'd learned that he was a brand-new father. As the navigator, the responsibility fell to him to bring the ten-man crew back safely. In retrospect, it's not terribly surprising that fate and the powers that be had selected Tom to be the navigator. He was the only one aboard the *I've Had It* who possessed a college degree. He was also the oldest man on the plane. He was twenty-three years old.

AS THE NAVIGATOR, THE RESPONSIBILITY FELL TO HIM TO BRING THE TEN-MAN CREW BACK SAFELY.

After each successful mission it was the custom to paint a small bomb on the side of the plane; in the rare instance of shooting down an enemy plane, a swastika was painted. When one incoming crew, however, accidentally hit a British runway maintenance worker, a small teacup was painted on the side of the plane, practically engendering an international incident.

Tom was a hero in what he still refers to as "the last good war." For his efforts, he received the Distinguished Flying Cross and the Air Medal with three Oak Leaf clusters and the heartfelt gratitude of his crew. Yet the commanding officer's first words to Tom and his young compatriots had not been wrong. The CO had told them to look at the man on their left and to look at the man on their right. "When you return," he'd said, "they will not be here." This dire prophecy

'Scuse Me While I Whip This Out   **191**

proved to be almost correct. The Mighty Eighth suffered a grievous attrition rate during the height of the war.

After the war Tom and Min settled in Houston, where Tom pioneered community action programs and Min became one of the first speech therapists in the Houston public schools. In the late fifties they moved to Austin, where Tom was a professor of educational psychology at the University of Texas. It was

"THE LAST TIME EVERYTHING WAS ALL RIGHT WAS AUGUST 14, 1945." THAT WAS THE DAY JAPAN SURRENDERED.

in 1953, however, that my parents made possibly their greatest contribution to children far and wide by opening Echo Hill Ranch. My mother passed away in 1985, but Tom, known as Uncle Tom to the kids, still runs the camp.

Like most true war heroes, Tom rarely talks about the war. My sister, Marcie, once saw Tom sitting alone in a darkened room and asked, "Is everything all right, Father?" To this Tom replied, "The last time everything was all right was August 14, 1945." That was the day Japan surrendered.

On a recent trip to O'Hare Airport in Chicago, I commandeered a limo and drove through the area where Tom had grown up. There were slums and suburbs and Starbucks, and the trolley cars and the clotheslines and the peddler with his horse and cart were gone. *"Kartofel,"* I said to the limo driver, but he just looked straight ahead. Either he wasn't Polish or he didn't want any potatoes.

Today Tom lives in Austin with his new wife, Edythe Kruger, and his two dogs, Sam and Perky. He has three children and three grandchildren. He eats lunch at the Frisco and still plays tennis with his old pals. He did not, as he contends, teach me everything I know. Only almost everything. He taught me ten-

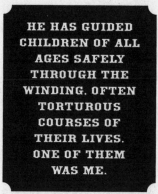

HE HAS GUIDED CHILDREN OF ALL AGES SAFELY THROUGH THE WINDING. OFTEN TORTUROUS COURSES OF THEIR LIVES. ONE OF THEM WAS ME.

nis. He taught me chess. He taught me how to belch. He taught me to always stand up for the underdog. He taught me the importance of treating children like adults and adults like children. He is a significant American because by his example, his spirit, and his unseen hand, he has guided children of all ages safely through the winding, often torturous courses of their lives. One of them was me.

Tom's war is long over. Indeed, the whole era seems gone like the crews who never came home, lost forever among the saltshaker stars. And yet, when the future may look its darkest, there sometimes occurs an oddly comforting moment when, with awkward grace, the shadow of a silver plane flies inexplicably close to my heart. One more mission for the navigator.

BOOKS BY KINKY FRIEDMAN

"The world's funniest, bawdiest, and most politically incorrect country-music singer turned writer."
—*New York Times Book Review*

'Scuse Me While I Whip This Out

Reflections on Country Singers, Presidents, and Other Troublemakers

ISBN 0-06-053976-3 (paperback)

"We don't presume to be able to explain the outlaw writer Kinky Friedman; we prefer to let him explain himself." —*Dallas Observer*

Kill Two Birds & Get Stoned

A Novel

A *New York Times* Notable Book and National Bestseller

ISBN 0-06-093528-6 (paperback)

A hilarious and provocative satire of the haves and have-nots, corporate sprawl, and the art and business of writing in this story of a novelist's life transformed by a pair of millennial merry pranksters.

"Great stuff. . . . Flashes of brilliance and laugh-out-loud observations." —*Denver Rocky Mountain News*

Kinky Friedman's Guide to Texas Etiquette

Or How to Get to Heaven or Hell Without Going Through Dallas–Fort Worth

ISBN 0-06-093535-9 (paperback)

As the oldest living Jew in Texas who doesn't own any real estate, Kinky considers it his duty to educate Texans and non-Texans alike on the customs and habits of his native state. You'll never look at Texas the same way again!

Want to receive notice of events and new books by Kinky Friedman?
Sign up for Kinky Friedman's AuthorTracker at www.AuthorTracker.com

---

Available wherever books are sold, or call 1-800-331-3761 to order.